Computational Intelligence for Wireless Sensor Networks

Computational Intelligence for Wireless Sensor Networks

Principles and Applications

Edited by

Sandip Kumar Chaurasiya

Joydeep Dutta

Arindam Biswas

Gorachand Dutta

Mrinal Kanti Sarkar

CRC Press is an imprint of the
Taylor & Francis Group, an **informa** business
A CHAPMAN & HALL BOOK

First edition published [2023]
by CRC Press
6000 Broken Sound Parkway NW, Suite 300, Boca Raton, FL 33487-2742
and by CRC Press

4 Park Square, Milton Park, Abingdon, Oxon, OX14 4RN

CRC Press is an imprint of Taylor & Francis Group, LLC

Library of Congress Cataloging-in-Publication Data

Library of Congress Cataloging-in-Publication Data
Names: Chaurasiya, Sandip Kumar, editor.
Title: Computational intelligence for wireless sensor networks : principles and applications / edited by Sandip Kumar Chaurasiya, Joydeep Dutta, Arindam Biswas, Gorachand Dutta, Mrinal Kanti Sarkar.
Description: First edition. | Boca Raton, FL : CRC Press, [2022] | Includes bibliographical references and index. | Identifiers: LCCN 2021059286 (print) | LCCN 2021059287 (ebook) | ISBN 9780367608880 (hbk) | ISBN 9780367608897 (pbk) | ISBN 9781003102397 (ebk)
Subjects: LCSH: Wireless sensor networks.
Classification: LCC TK7872.D48 C645 2022 (print) | LCC TK7872.D48 (ebook) | DDC 006.2/5--dc23/eng/20220124
LC record available at https://lccn.loc.gov/2021059286
LC ebook record available at https://lccn.loc.gov/2021059287

ISBN: 978-0-367-60888-0 (hbk)
ISBN: 978-0-367-60889-7 (pbk)
ISBN: 978-1-003-10239-7 (ebk)

DOI: 10.1201/9781003102397

Typeset in Palatino
by KnowledgeWorks Global Ltd.

Contents

Preface

Wireless Sensor Networks (WSNs) are networks of tiny electromechanical devices, more precisely called sensor nodes. These sensor nodes are equipped with the ability to sense and measure their surroundings. The popularity of such networks can easily be perceived through a wide variety of applications such as habitat monitoring, health monitoring, security and surveillance, civil structure monitoring, precision agriculture, animal tracking, and industrial applications. However, these networks suffer from several constraints such as storage and computational limitations, limited power resources, and limited transceiving capabilities. The limitations mentioned above result in many challenges such as deployment and localization, energy-efficient data gathering and routing, data fusion, security, task scheduling, etc.

Over the past few years, computational intelligence (CI) has emerged as an effective tool to address these challenges. Computational intelligence refers to a set of adaptive techniques facilitating intelligent behavior in complex and dynamically changing environments like wireless sensor networks (WSNs). The elements of learning, adaptation, and evolution are integrated to create an intelligent system via CI; thus, CI enables WSNs to exhibit autonomous behavior in a rapidly changing environment and provides robustness against the above-cited challenges.

This work attempts to bring the learning of technologies such as CI and WSN to foster strong collaboration between them. The following chapters thoroughly discuss WSNs and sensor-enabled technologies along with their respective properties and challenges. Furthermore, the present work will discuss various CI techniques such as fuzzy computing, evolutionary computing, reinforcement learning, artificial intelligence, swarm intelligence, and their respective applications in wireless sensor networks and sensor-enabled technologies in greater depth.

About the Authors

Sandip K. Chaurasiya received his M.E. degree in Computer Science and Technology from the Indian Institute of Engineering Science and Technology Shibpur, Howrah, West Bengal, India, in June 2010. He is working as an Assistant Professor in the University of Petroleum and Energy Studies, Dehradun, Uttarakhand, India. His research interests include mobile computing, wireless sensor networks, and the internet of things. He has authored a book, *A Complete Guide to Wireless Sensor Networks: From Inception to Current Trends* (2019), and contributed many technical papers in various journals and conferences. He has reviewed a number of research articles in several reputed international conferences/journals. He is an active member of the Institution of Electronics & Communication Engineers (India), New Delhi.

Dr. Joydeep Dutta completed his Ph.D. degree from the Department of Computer Science and Engineering, NIT Durgapur in 2020. He completed his M.E. in Computer Science and Engineering from the West Bengal University of Technology in 2008. Dr. Dutta was also awarded the bronze medal for his first-class 3rd position in the university. Dr. Dutta qualified with the UGC NET in Computer Science and Applications in 2013. He is currently working as an Assistant Professor in the Department of Computer Science of Kazi Nazrul University, Asansol, India. He has more than nine years of teaching experience in several reputed engineering colleges of West Bengal and two years of industrial experience in MNCs such as Wipro Technologies and Oracle Financial Services Software Ltd. His current research interests include network optimization problems, artificial intelligence, evolutionary algorithms, combinatorial optimization etc. He has several high-impact SCI-indexed publications like *Neural Computing and Applications, Transport,* etc. Dr. Dutta has acted as a reviewer of various reputed journals and conferences. He has chaired, organized, and presented papers at seminars and conferences of international repute.

Dr. Arindam Biswas was born in West Bengal, India, in 1984. He received his M-Tech degree in Radio Physics and Electronics from the University of Calcutta, India, in 2010 and his Ph.D. from NIT Durgapur in 2013. He was a Post-Doctoral Researcher at Pusan National University, South Korea, with prestigious the BK21PLUS Fellowship, Republic of Korea. He was Visiting Professor at the Research Institute of Electronics, Shizouka University, Japan. He has been selected for the IE(I) Young Engineer Award: 2019–20 in Electronics & Telecommunication Engineering, Institute of Engineers, India. Dr. Biswas has 11 years of experience in teaching research and administration. Presently Dr. Biswas is working as an Assistant Professor in the School of Mines and Metallurgy at Kazi Nazrul University, Asansol, WB, India. He has 48 technical papers in various journals and 30 conference proceedings and six books, one edited volume and one book chapter with international repute. Dr. Biswas has received a research grant from the Science and Engineering Research Board, Government of India, under the Early Career Research Scheme for research in Terahertz-based GaN Source. He has also received a research grant from the Centre of Biomedical Engineering, Tokyo Medical and Dental University, in association with RIE, Shizouka University, Japan, for the study of biomedical Thz Imaging based on WBG semiconductor IMPATT Source. Presently Dr. Biswas is serving as an Associate Editor of Cluster Computing, Springer (SCI Indexed) and as a guest editor of *Nanoscience and Nanotechnology-Asia* (Scopus Indexed), *Recent Patents in Material Science* (Scopus Indexed), Bentham Science Publisher. Dr. Biswas has advised four Ph.D. students in different topics related to applied optics and high-frequency semiconductor devices. He has organized and chaired various international conferences in India and abroad. His research interests are in carrier transport in low-dimensional systems and electronic devices, non-linear optical communication, and THz semiconductor sources. Dr. Biswas has acted as reviewer for reputed journals and is a member of the Institute of Engineers (India) and Regular Fellow of Optical Society of India (India).

Dr. Gorachand Dutta is an Assistant Professor at the School of Medical Science and Technology, IIT Khragpur. He received his M.Sc. degree in Chemistry from the Indian Institute of Technology, Guwahati, India. His research interests include the design and characterization of portable biosensors, biodevices and sensor interfaces for miniaturized systems and biomedical applications for point-of-care testing. He received his Ph.D. in Biosensors and Electrochemistry from Pusan National University, South Korea, where he developed a different class of electrochemical sensors and studied the electrochemical properties of gold-, platinum-, and palladium-based metal electrodes. He completed his post-doctoral fellowships in the Department of Mechanical Engineering, Michigan State University, USA; the Department of Chemistry, Pusan National University, South Korea; and the Centre for Biosensors, Bioelectronics and Biodevices at the University of Bath, UK. During his research tenure in the USA and South Korea, Dr. Dutta invented an enzyme-free, disposable miniaturized immunosensor chip using micropatterned electrodes and a wash-free method for the development of mobile phone-based platforms for fast and simple point-of-care testing of infectious and metabolic disease biomarkers. He has expertise on label-free multichannel electrochemical biosensors, electronically addressable biosensor arrays, aptamer- and DNA-based sensors, and surface bio-functionalization.

Dr. Mrinal Kanti Sarkar (Senior Member, IEEE) received his Ph.D. degree in Electrical Engineering from the National Institute of Technology Durgapur, India, in 2015. Presently he is an Assistant Professor of the Electrical Engineering Department at NIT Manipur, India. Dr. Sarkar served as Departmental Head from January 2017 to January 2019. Presently he is serving as Controller of Examinations in addition to his normal duties. His research interests include magnetic levitation systems, DC-DC converters, optimal control, sliding mode control, event-triggered control and networked-control systems.

Contributors

Arjun Arora
Cybernetics Clusters, School of Computer Science
University of Petroleum and Energy Studies
Dehradun, Uttarakhand, India

Geetansh Atrey
School of Computer Engineering
KIIT Deemed to be University
Bhubaneswar, India

Gaytri Bakshi
Cybernetics Clusters, School of Computer Science
University of Petroleum and Energy Studies
Dehradun, Uttarakhand, India

Rajib Banerjee
Dr. B. C. Roy Engineering College
Durgapur, West Bengal, India

Sandip K. Chaurasiya
Cybernetics Clusters, School of Computer Science
University of Petroleum and Energy Studies
Dehradun, Uttarakhand, India

Debasis Giri
Department of Information Technology
Maulana Abul Kalam Azad University of Technology
West Bengal, India

Suman Ghosh
RACE
United Kingdom Atomic Energy Authority, UK

Tanmoy Maitra
School of Computer Engineering
KIIT Deemed to be University
Bhubaneswar, India

P. K. Paul
Department of CIS,
Information Scientist (Offg.),
Raiganj University (RGU)
West Bengal, India

Himanshu Sahu
Cybernetics Clusters, School of Computer Science
University of Petroleum and Energy Studies
Dehradun, Uttarakhand, India

T. Santosh
School of Business
Woxsen University
Hyderabad, India

Arup Sarkar
School of Computer Engineering
KIIT Deemed to be University
Bhubaneswar, India

Ritwik Saurabh
School of Computer Engineering
KIIT Deemed to be University
Bhubaneswar, India

Dhirendra Kumar Sharma
Cybernetics Clusters, School of Computer Science
University of Petroleum and Energy Studies
Dehradun, Uttarakhand, India

Hitesh Kumar Sharma
Cybernetics Clusters, School of Computer Science
University of Petroleum and Energy Studies
Dehradun, Uttarakhand, India

Shiwangi Singh
School of Computer Engineering
KIIT Deemed to be University
Bhubaneswar, India

Rohit Srivastava
Cybernetics Clusters, School of Computer Science
University of Petroleum and Energy Studies
Dehradun, Uttarakhand, India

1

Wireless Sensor Network (WSN) Vis-à-Vis Internet of Things (IoT) Foundation and Emergence

P. K. Paul

CONTENTS

DOI: 10.1201/9781003102397-1

1.1 Introduction

Wireless sensor networks are autonomous sensors that are responsible for the collection, cooperation, and monitoring of physical as well as environmental quantities and qualities and allied interests, viz. temperature, sound, vibration, pressure, pollutants and so on [1–3]. Initially wireless sensor networks were projected for defense and military applications; gradually they are becoming important tools for societal applications and civilian applications in diverse activities such as

- healthcare and monitoring
- transportation and traffic management
- environment and ecological monitoring and management
- industrial automation, and so on.

Two standard communication protocols have been proposed in past few years: Wireless HART and ISA100.11a; in this regard, the HCF consortium and ISA Association play a leading role. These tools are designed for process and monitoring as well as control. Each node in the wireless sensor network is equipped with a sensor, a small microcontroller responsible for analog-to-digital signal conversion, a computational unit, and storage systems [3–5]. The gradual development of the internet emerged as the internet of people (IoP), and in recent past the internet of things concept has become widely recognized and popular. It is further expected that IoT will add about 75 billion things and items through the internet by the end of 2025 [6].

The shift in electronics products includes radio frequency identification (RFID), mobile devices, and wireless sensors etc. Here IPv6, IPv6 over low power wireless personal area networks sensors standards are playing a leading role. Here M2M (machine-to-machine communication) is also an important concept. Integration and applications of wireless sensor networks in the IoT is worthy and important. In the internet of things, machines and electronic devices become more and more connected with other things or objects. To communicate autonomously with the help of internet, the emergence of IoT is noticeable and acceptable. In 2019 the total number of IoT devices reached 26.66 billion, and it is worthwhile to note that every second 127 new IoT devices are connected with the web. Billions of things are connected to the internet, and they are dedicated to data generation. In IoT, M2M is the main communication standard between the internet of things [7–9].

1.2 Objectives

The present work has the following (but not limited to) objectives:

- To know about the basics of wireless sensor networks with reference to their foundation, features, and characteristics.
- To learn about the role and importance of the wireless sensor network with reference to the present context.

- To know about the basics of cloud computing in relation to wireless sensor networks.
- To learn about the basics of the internet of things with reference to the foundations, features and characteristics.
- To learn about the growing applications of the internet of things in different sectors and areas.
- To know about the role and need of wireless sensor networks in the internet of things context.

1.3 Methods

This work is theoretical in nature and deals with various aims and objective as discussed. The topic of the work "Wireless Sensor Network and its utilizations in internet of things: *Foundation and Emerging Trends*" is theoretical in nature; therefore a review of literature played a leading role in the formation of the paper. Secondary sources were initially reviewed, and thereafter various primary sources were also exploited for doing the research work. Various websites of companies offering IoT and WSN services were also reviewed and analyzed to get a concise picture of the topic.

1.4 Wireless Sensor Networks: Foundations

Collection of nodes make up wireless sensor networks. Such nodes are individual small computers. WSNs form centralized network systems and enable multi-functionality while being wireless in nature. Wireless sensor networks may have a predefined goal; having a centralized and synchronized structure, they follow a certain topology like linear, star, mesh etc. Limited broadcast range in a wireless sensor network normally is 30 meters. Wireless sensor networks normally having following steps:

- collecting data
- processing data
- packaging data
- transferring data

Wireless sensor networks are similar to wireless ad hoc networks as they also collect the data with wireless support, and it is important to note that wireless sensor networks are the kind of sensors that are autonomous in nature and are responsible for the physical or environmental conditions including the temperature, sound pressure and so on [9–11]. Here data basically are moved by the cooperation of the network to the main location. Modern networks may be bi-directional, collecting data from the distributed sensors and also dedicated in controlling the sensor activities. Military applications like battlefield surveillance were the main reason for the development of the WSN, but today WSNs are widely used in many industrial as well as consumer applications that are built of "nodes;"

the number may be a few to hundreds or even thousands, where each node is connected to one sensor. Such sensor network nodes may have various types of parts, viz. internal antenna, external antenna, microcontroller, and an electronic circuit for the connectivity of the sensors and an energy source, battery, or other energy-harvesting system. It is worth noting that the sensor node size may be different, and further the cost of sensor nodes is also variable depending upon the brand and country. As far as the topology is concerned WSNs may be prepared with a simple star network or even a multi-hop wireless mesh-based network. The propagation technique of hops of the network may be routing or flooding [12–14]. As far as characteristics are concerned the following can be consider as important and valuable:

- Power consumption constraints can be noted for the nodes using batteries or energy harvesting systems.
- Ability to cope with node failures is important.
- In certain case the mobility of nodes can be noted.
- Heterogeneity and homogeneity of the nodes can be important.
- Scalability is an important feature, emphasizing large scale of deployment.
- WSNs are very effective to use.
- Cross-layer optimization is important and can be noted.

The internet of things is an important technology that offers scalable and mobile computing services. Various technologies play an important role for the further development of IoT; among these are ubiquitous and pervasive computing. These technologies are helping in the development of wireless systems by producing, and consuming RFID and mobile computing-based services. RFID is a useful early example of IoT applications where WSNs are also effectively started a few years back; and among the examples are goods, cars, wearable sensors, etc. RFID basically uses less energy than mobile and handheld devices. With the support of cloud computing, the capabilities of these devices will be further boosted in respect of the storage and other infrastructure-based services [15–17].

1.5 Wireless Sensors Network: Emergence and Basic Applications

Wireless sensor networks run in a bi-directional fashion and it are wirelessly connected with networks of various kind of sensors. These sensors are dedicated to collection of data on different aspects, viz. temperature, humidity, speed, etc. Communication is done on a multi-hop basis, and each sensor is dedicated to perform the defined task. Wireless sensor networks follow OSI architecture and model and have five layers and three cross layers. For the active operation it is essential to have five layers:

- application layer
- transportation layer
- network layer
- data link layer
- physical layer

Wireless sensor networks are able to offer various advantages and benefits; these include the following:

- Network arrangements can be brought in without involvement of immovable infrastructure.
- WSNs are effective IN non-reachable places, viz. over the sea, rural areas, deep forests, hills, and mountain areas.
- They are very flexible in casual situations.
- The cost of the tools and technologies, especially the execution pricing, in wireless sensor network is inexpensive.
- Wireless sensor networks avoid wiring.
- Wireless sensor networks may also able accommodate many kind of devices at any time.
- With the help of centralized monitoring, wireless sensor networks become worthwhile.

Wireless sensor networks offer various types of applications that come with the comfortable, effective, and smart life. Energy saving and minimal noise are other benefits of WSNs [18–20]. Wireless sensor networks contribute effectively in the development of less costly atmospheric monitoring system while reducing the pollution, and hence, facilitating the healthcare benefits too. In wireless sensor networks application-based communication requires sensors and different kinds of server connectivity. Wireless sensor networks basically function by the use of three main access technology architectures. Various kinds of sensors are used, viz. low sampling rate, seismic, magnetic, thermal, visual, infrared, radar, and so on based on situation and usefulness. Sensor nodes are very useful in constant sensing, including in event ID, event detection, and so on. The application of wireless sensor networks initially was for military and defense-related purposes, and gradually it has been updated in other areas:

- medical and healthcare sectors and applications
- environmental and ecological applications
- home and living applications
- organizational and commercial applications
- area monitoring and management
- earthquake sensing and similar activities
- air pollution monitoring and management
- forest fire detection
- landslide detection and similar activities
- water quality and ocean monitoring with management

Wireless sensor networks have longer range than that of the basic sensor networks, and WSNs are based on communication in a peer-to-peer basis. The maximum number of wireless sensor networks is based on IEEE 802.15.4 standard and is connected with the physical and medium access control (MAC) layer of low rate-wireless personal area networks (LR-WPANs). In effective WSNs, data mining techniques are also used to extract enormous amount of data. In WSNs the sensors are always active and reduce the energy consumption according to the need [21, 22]. The sensors of WSNs collect the data frequently and store them in the cloud or database accordingly.

1.6 Internet of Things: Foundation and Emergence

Entrepreneur Kevin Ashton initially coined the term internet of things, or IoT, in 1999, and gradually it has become an important field of practice in information technology. Though it is important to note that the technology rapidly emerged and grew in past decade. In 2011 Gartner reported that IoT was one of the latest technologies; thereafter the growth of the IoT was noticeable. In the recent past, the abbreviation IoT became popular as the emerging technology is applied to various kinds of objects. Industrial machines become wearable devices with the help of IoT. Ashton played a leading role in IoT development under the auspices of Auto-ID Center at Massachusetts Institute of Technology. Many experts initially called it "embedded internet," and today it is embedded in our daily lives. There is no universally accepted and perfect definition of the internet of things; various experts viz. scientist, academician, researchers, practitioners, engineers, developers and industry persons define it differently [23, 24].

The internet of things is an open and comprehensive network of objects that are capable of decision making, data and resource sharing, intelligent computing, and manufacturing, connecting intelligent devices with sensors and the internet. The internet of things is maturing and getting more sophisticated and it is making the world more completely a global village with various kind of connected objects. Further IoT provides services anytime and anywhere using the internet and sensors and it is considered a global network. This network further is responsible for the human-to-human, human-to-object or thing, and thing-to-thing communication. With the help of IoT unique identity for every object can become possible and may be connected and worked accordingly. With IoT more can be communicated in an intelligent fashion than ever before and "being connected" becomes possible. Various electronic devices can be connected, viz. servers, computers, smart phones etc. The internet of things is integrated with sensors and actuators embedded in physical objects, and it also linked by wired and wireless networks using IP. Due to the increasing components in IoT (refer to Figure 1.1) the services are changing and improving day by day.

IoT has become an important technology connecting physical and digital components. The components of the IoT are able to transmit data without human mediators and each component has a unique identifier (UID). IoT applications can be following types.

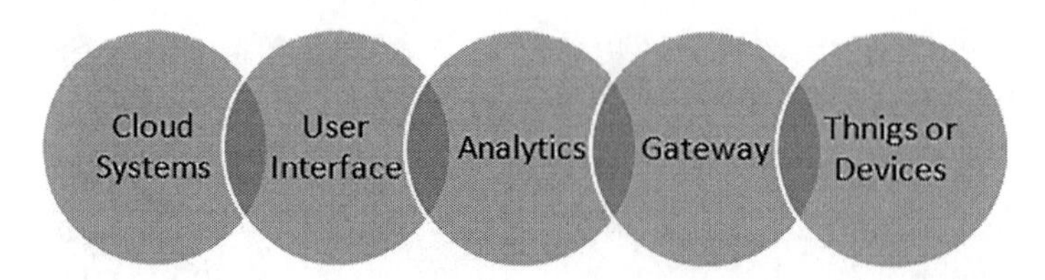

FIGURE 1.1
Major components of IoT.

1.6.1 Consumer IoT

Consumer-based IoT is dedicated to the consumer only, viz. light fixtures, home appliances, old-age services, nursing services, entertainment-related activities, and so on [25, 26].

1.6.2 Commercial IoT

Applications of the internet of things is very effective and important in different commercial activities, including healthcare, transportation, manufacturing devices, smart pacemakers, office- or establishment-based monitoring systems, and in the emerging vehicle-to-vehicle communication, and so on.

1.6.3 Industrial Internet of Things (IIoT)

IoT applications in industry launched the industrial internet of things, dedicated to making industries more advanced and smarter, viz. digital control systems, statistical evaluation, smart and digital agriculture, and design and development of industrial products.

1.6.4 Infrastructure IoT

Applications of the IoT are increasing day by day in the infrastructure sector, enabling the connectivity of smart cities and organizations in building of healthy and sophisticated infrastructure with the support of the sensors, management systems, and other intelligent electronic system support [27, 28].

1.6.5 Internet of Military Things (IoMT)

The utilization of the IoT technologies in the areas of defense as well as military sectors has also been increasing, viz. robots for surveillance and human-wearable biometrics for combat and so on.

These are the basic types of IoT, and such terminology is increasing. In 2018 about 7 billion IoT devices existed, and gradually the number has increased to 26.66 billion and it is expected to grow to about 75 billion by the end of 2025. IoT applications are noticeable in different sectors, viz. business, commercial ventures, and industries; agricultural, horticultural, and environmental sciences; education, teaching and research, government and other management-related organizations; healthcare and medical systems; transportation and tourism; manufacturing organization systems, etc. In diverse areas IoT applications are possible with wireless internet and embedded sensors with various technology. Due to the nature and uses of the IoT in different sectors, the IoT architecture (refer to Figure 1.2) is also changing gradually and thus IoT can be consider as additionally as follows:

- The Internet of things can be treated as valuable in the information technology age for the development of smarter information solutions with allied technologies.
- Cloud computing, data analytics, usability engineering, human computer interaction (HCI) can be considered a valuable technology in respect of IoT support.
- It is sensor dependent, and various grids are connected; further most of these are renewable [29, 30].
- Machine monitoring sensors are getting intelligent machine and systems development support for complete internet-based services and products.

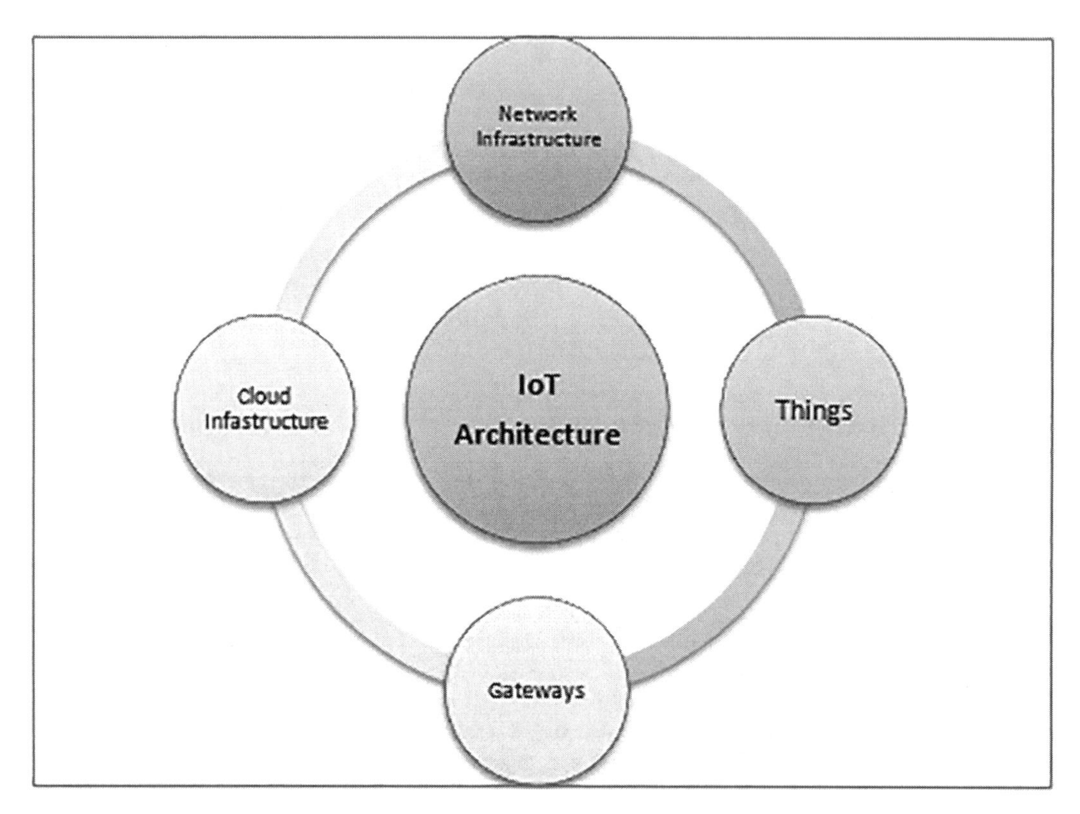

FIGURE 1.2
The IoT architecture at a glance.

1.7 IoT: Emerging Applications

The internet of things is an internet-based system, digitally connected and controlled; therefore it is applicable in wider areas, including in efficiency, safety, security, etc., in diverse areas such as oil and gas, education and training, manufacturing, transportation, tourism, agriculture, retail and hospitality sectors, etc.

The internet of things and its implementation is an important issue as high organizational cost can be applicable in the broad areas of IoT. In home automation, smart towns and cities, and other smart development, IoT is also very important and needed.

In developing the smart home, viz. air conditioning, lighting, heating, security etc., IoT plays a leading role. Smart devices such as iPhone or iOS are useful in this context, and the smart building is also an emerging concept and tool of IoT for reducing energy consumption. With the help of IoT, elderly individuals and people with disabilities can use assistive technology and voice control systems; therefore IoT applications are increasing.

Smarter healthcare is an integrated approach of IoT in healthcare and medical systems such as collection and analysis of data. Remote emergency services are a core part of IoT systems. The use of the pacemakers, smart beds, nursing informatics, medical facilities, and "m-health" (mobile health services) is very important [31, 32]. Different kinds of IoT-enabled social applications such as vehicle management, traffic control, electronic toll collection, and infrastructure development & management are very

FIGURE 1.3
Major using devices of IoT.

important and growing. Thus, IoT has evolved as a need of the hour in almost every aspect of our life.

In manufacturing IoT is important in identification, processing, and communication. It is also important for better manufacturing equipment and rapid manufacturing, proper supply chain management, and smart grid-based systems. The concept of the industrial internet of things is important but is possible only with proper support from all the stakeholders and institutions. Day by day the use of electronic products is rising, including various items, viz. refrigerators, air conditioning, simple computing devices, cameras, or even a simple blender. Figure 1.3 provides details.

1.8 WSN, Cloud Computing and Internet of Things

Cloud computing is an important concern in modern information technological tools; it is responsible for the design and development of virtual information technology systems, including software, applications, information technology infrastructure, operating systems, platforms, and so on. There are different types of cloud computing:

- public
- private
- hybrid

Public cloud computing basically uses internet services, which helps in offering cloud-based tools, techniques, and services [33, 34]. Higher bandwidth and sophisticated internet connections are an important concern. Private cloud computing is similar to the public offering but services are normally designed and developed in-house. There are various cloud computing services based on service models; the major ones are software as a service, platform as a service, security as a service, storage as a service, infrastructure as a service, and so on. Cloud computing is highly connected with the wireless sensor network as both are responsible for making IT systems remote and wireless. Cloud computing is significant in a developing virtual world. Similar to cloud computing another concept is also significant: big data, or analytics. The combination of wireless sensor networks and the internet of things is important since both are dedicated in wireless IT support. The mixture of IoT and cloud with the WSN enables the provision of sensor data or sensor event as a service over the internet. Therefore sensor data can be easily analyzed locally and everywhere with a proper system that connects to the wireless systems [35, 36]. WSN architecture is also advantageous in respect of mode flexibility, viz. task management, mobility management, power management, and so on. Refer to Figure 1.4 to learn about more about WSN architecture and core benefits.

Due to the importance and role of the wireless sensor networks and cloud computing (and partially IoT), sensing as a service and sensor event as a service has emerged. Sensor data basically is made available to the clients across the cloud and similar type of infrastructure. These combined technologies are significant for a large number of different applications and are increasing. Among the basic and core uses few important are the following.

1.8.1 Transport and Tourism Monitoring

There are various places in which systems of traffic control can be supported by WSN-integrated IoT: license plate recognition, toll way management, normal as well as emergency vehicle management, smarter traffic lights, emergency road management, etc. WSN-integrated IoT-based devices can collect data and store them in the cloud and make

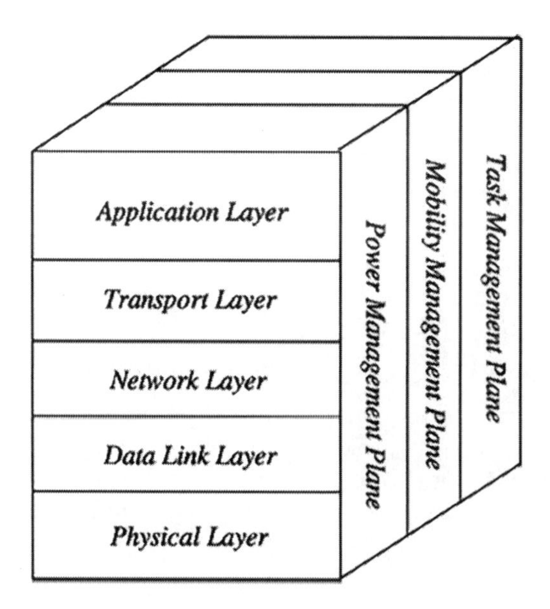

FIGURE 1.4
Layers vs. effectiveness of WSN.

transportation and traffic systems more effective and modern. In a variety of areas these collected data are useful, viz. in vehicle classification, collision avoidance systems, automatic toll gates, and others [37, 38].

1.8.2 Military Use

Wireless sensors networks initially were developed to solve problems of military and defense sector, and gradually it has been recognized as useful in other sectors. Initially it was for the smart dust concept. Since a wireless sensor network is internet based, therefore cloud computing uses are important to note.

1.8.3 Weather Forecasting

With the support of cloud computing and big data, wireless sensor network–based devices integrated with the IoT are able to make weather forecasting systems more useful and important. WSN-supported sensors basically collect numerous data that help in weather forecasting [17, 39, 40].

1.8.4 Healthcare

Sensors are effective and useful in data collection and highly important in healthcare and medical applications. Personal or individual devices, many of them portable, are increasing in healthcare sectors. These can be effective in collection of the data from the patient and helpful in decision-making. WSNs for real-time warnings for safer and more effective patient management are important. Furthermore, WSN and IoT are important in constant and remote and personalized patient care [18, 41, 42]. The following sections are noteworthy in respect of emerging and future applications of IoT, WSNs, and integrated approaches.

1.9 WSNs in Smarter IoT Systems: Emerging Applications and Trends

As far as technological scope is concerned a WSN is smaller than the internet of things. Wireless sensor networks are not directly connected with the internet but are mainly connected with the central node or a router. But the internet of things is directly connected with the internet or it is internet-based only. Therefore IoT can collect data from a wireless sensor network; a sensor can collect the data and can also store it in the cloud, where it can be utilized based on need and requirement. For example, the sensors are able to collect data viz. temperature, wind movement, etc., and data can be sent periodically to the internet, where a server can process the data and interpret accordingly. We are already aware that IoT is based on internet and sensor connections, but a WSN can be seen as a group of sensors, or in as sense, it is the big sensor. Therefore IoT exists at higher level then wireless sensor networks. Since a WSN may be considered as a part of IoT, therefore the function areas can be seen as similar. In a mesh network or similar, WSN can be used to gather data by a router. A wireless sensor network consists of a network of only wireless sensors, normally without any wired sensor.

The internet of things has unique identifiable embedded computing devices within the existing internet framework and offers sophisticated, advanced connectivity to the devices, systems, and services, and that may be beyond traditional machine-to-machine communication and based on different kind of protocols, applications etc. This helps in automation and also enables advanced smart grid types of applications. Things basically communicate with each other without interaction with a human or system. IoT from core sense has three basic components:

- sensors
- actuators
- connectivity devices

There are broad agreements among the connectivity between the devices' security and privacy risks, and all these are major issues in WSN. Wireless sensor networks integrated with IoT may be helpful in the following areas:

- enabling authorized access
- preventing the misuse of personal information
- facilitating the reduction of attacks on a system
- mitigating the safety risks

Security is an important concern. In the information technology field there are tools, devices, and components that pose possible threats. Wireless sensor networks may have different kind of security-related issue. The internet of things also has various security-related issues, and organizations have to deal with such security-related concern very carefully [5, 43, 44]. All the components of the IoT should be considered of prime importance in this aspect of security. (Refer to Figure 1.5 for more details.)

Technologies like radio frequency identification and wireless sensor networks, which are integrated with the IoT, help in decision making and various applications. They are increasing in wide areas apart from the previously mentioned.

1.9.1 Smart Home and Residential

In home automation and advanced management of the home and residential complexes WSN and IoT-based tools and technologies are emerging rapidly for a wide range of areas, viz. utility meters, energy and power supply management, water supply and water leaks, lift and door systems, remote operation and monitoring using WSN/IoT based systems, and so on.

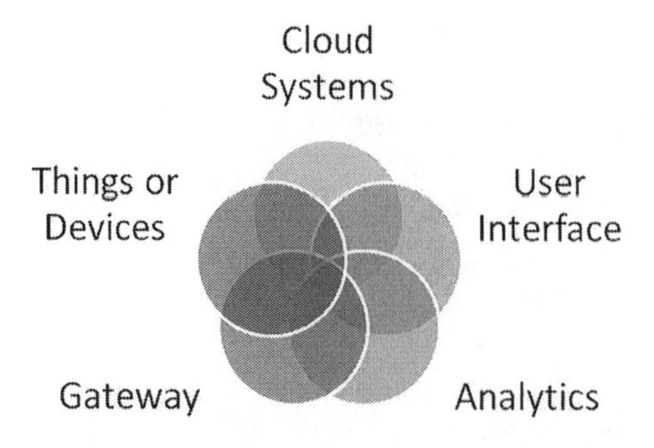

FIGURE 1.5
Major components of IoT where security aspects are important and connected.

1.9.2 Smart Environmental Management and Monitoring

In various environment and ecology related areas IoT and wireless sensor networks are useful, such as in flood management, fire management, earthquake management, disaster management, and air pollution management.

1.9.3 Smart Hospitals and Healthcare

In future hospitals and healthcare organizations there will be more automation and much technological involvement. There are numerous areas where wireless sensor networks, RFID, or IoT can be connected, viz. in managing heart rate, blood pressure, temperature of the patient, and the room. Apart from these uses in hospital management, IoT and WSN integrated technologies are highly important and required for remote operations, electronic device management, medical emergencies, and medical transportation. Drone ambulances can fly to the scene with the emergency kit and some specially designed assistance drones. IoT-based automated information system design and development are also noteworthy and very important [19, 45].

1.9.4 Smart Transportation and Tourism

In transportation and allied fields IoT and WSN-based tools and technologies are noteworthy and important. There are traditional and diverse areas where IoT and WSN are important to use, viz. real-time traffic and public transportation systems using information sharing, intelligent and automated traffic control systems, electric vehicle management, automated charging facilities, short-range communication.

a. *Proper navigation and safety of transportation*—In different types of vehicles, such as cars, buses, and trains, as well as routes and roads the use of sensors and actuators is valuable for the drivers as well passengers regarding better navigation and safety.
b. *Road planning and route optimization*—More accurate traffic information, including road patterns, planning, warning messages on climate conditions, and traffic jams, can be possible with IoT and WSN-based systems.

1.9.5 Real-Time Management and Decision Making

Time is an important aspect for all kinds of organizations, institutions, and individuals, and everywhere time is an important concern. IoT and wireless sensor network-based systems would be an important example in regard to time management. Sensors collect data and provide the same to the user, and based on this, real-time management and decision making become possible. Therefore in organizational support and development IoT and similar system such as WSN are noteworthy [37, 46].

1.9.6 Smart Gardening and Farming

IoT and WSN based systems are worthy and important in advanced and intelligent farming, including vertical farming also. Among the areas where IoT and integrated WSN are useful are automated lighting, humidity and temperature control, proper moisture identification and arrangements, etc.

1.9.7 Security and Privacy in Different Stakeholders

In proper and solid security privacy management of different products, organizations, and institutions IoT and WSN-based devices can be helpful in achieving security methods and in finding the existing trust relationship between the devices and server [2, 20, 47].

1.10 Future Potentialities

The internet of things and wireless sensor networks are very important and valuable for the development of organizations and institutions of different kind, including for profit and nonprofit, small and large. Proper planning and interest in this regard is important for different stakeholders, including government, institutions, policy makers, and so on. As the technology grows and it comes with aspects of cost; therefore proper funding is also important in the making of the intelligent world using different technologies, which include IoT, wireless sensor networks in association with allied technologies, viz. cloud computing, big data, and so on.

1.11 Conclusion

Wireless sensor networks (WSNs) are responsible for sensing, monitoring and controlling options of data. From the military applications it has now developed valuable environmental applications, lifestyle and citizen science, and industrial applications. Environmental applications, viz. forest management, disaster management, and air- and water-quality management can be considered important with WSNs. A wireless sensor network is combined with a network of nodes that collaborate and monitor the surrounding environment. An interaction between people and nodes is fundamental in wireless sensor networks. The connections, requirement, and affiliation of the internet of things can be considered important in developing a perfect combination of the IoT and WSN. The growing applications in diverse sectors regarding the WSN and IoT are a significant move in building an intelligent and digital society.

References

1. Aazam, M., Zeadally, S., & Harras, K. A. (2018). Offloading in fog computing for IoT: review, enabling technologies, and research opportunities. *Future Generation Computer Systems, 87,* 278–289.
2. Kaiwartya, O., Abdullah, A. H., Cao, Y., Lloret, J., Kumar, S., Shah, R. R., ... & Prakash, S. (2017). Virtualization in wireless sensor networks: fault tolerant embedding for internet of things. *IEEE Internet of Things Journal, 5*(2), 571–580.
3. Kulkarni, A., & Sathe, S. (2014). Healthcare applications of the internet of things: a review. *International Journal of Computer Science and Information Technologies, 5*(5), 6229–6232.
4. Behera, T. M., Mohapatra, S. K., Samal, U. C., Khan, M. S., Daneshmand, M., & Gandomi, A. H. (2019). Residual energy-based cluster-head selection in WSNs for IoT application. *IEEE Internet of Things Journal, 6*(3), 5132–5139.

5. Jin, J., Gubbi, J., Marusic, S., & Palaniswami, M (2014). An information framework for creating a smart city through internet of things. *IEEE Internet of Things Journal, 1*(2), 112–121.
6. Alyasiri, H., Clark, J. A., Malik, A., & Frein, R. D. (2021). Grammatical evolution for detecting cyberattacks in internet of things environments. *IEEE ICCCN at Athens, Greece*, July 2021, pp. 1–6.
7. Deif, D., & Gadallah, Y. (2017). A comprehensive wireless sensor network reliability metric for critical internet of things applications. *EURASIP Journal on Wireless Communications and Networking, 2017*(1), 145
8. Shen, J., Wang, A., Wang, C., Hung, P. C., & Lai, C. F. (2017). An efficient centroid-based routing protocol for energy management in WSN-assisted IoT. *IEEE Access, 5*, 18469–18479.
9. Xie, S., & Wang, Y. (2014). Construction of tree network with limited delivery latency in homogeneous wireless sensor networks. *Wireless Personal Communications, 78*(1), 231–246.
10. Fadel, E., Gungor, V. C., Nassef, L., Akkari, N., Malik, M. A., Almasri, S., & Akyildiz, I. F. (2015). A survey on wireless sensor networks for smart grid. *Computer Communications, 71*, 22–33.
11. Li, X., Niu, J., Kumari, S., Wu, F., Sangaiah, A. K., & Choo, K. K. R. (2018). A three-factor anonymous authentication scheme for wireless sensor networks in internet of things environments. *Journal of Network and Computer Applications, 103*, 194–204.
12. Aranzazu-Suescun, C., & Cardei, M. (2017). Distributed algorithms for event reporting in mobile-sink WSNs for Internet of Things. *Tsinghua Science and Technology, 22*(4), 413–426.
13. Bawany, N.Z., & Shamsi, J.A. (2015). Smart city architecture: vision and challenges. *International Journal of Advanced Computer Science and Applications, 6*(11), 246–255.
14. Kumar, S., & Chaurasiya, V. K. (2018). A strategy for elimination of data redundancy in Internet of Things (IoT) based wireless sensor network (WSN). *IEEE Systems Journal, 13*(2), 1650–1657.
15. Elijah, O., Rahman, T. A., Orikumhi, I., Leow, C. Y., & Hindia, M. N. (2018). An overview of Internet of Things (IoT) and data analytics in agriculture: benefits and challenges. *IEEE Internet of Things Journal, 5*(5), 3758–3773.
16. Preeth, S. S. L., Dhanalakshmi, R., Kumar, R., & Shakeel, P. M. (2018). An adaptive fuzzy rule based energy efficient clustering and immune-inspired routing protocol for WSN-assisted IoT system. *Journal of Ambient Intelligence and Humanized Computing*, 1–13.
17. Xu, G., Shen, W., & Wang, X. (2014). Applications of wireless sensor networks in marine environment monitoring: a survey. *Sensors, 14*(9), 16932–16954.
18. Elappila, M., Chinara, S., & Parhi, D. R. (2018). Survivable path routing in WSN for IoT applications. *Pervasive and Mobile Computing, 43*, 49–63.
19. Jaladi, A. R., Khithani, K., Pawar, P., Malvi, K., & Sahoo, G. (2017). Environmental monitoring using wireless sensor networks (WSN) based on IOT. *International Research Journal of Engineering and Technology, 4*(1), 1371–1378.
20. Paul, P., Bhuimali, A., & Aithal, P. S. (2017). Emerging internet services vis-à-vis development: a theoretical overview. *International Journal on Recent Researches in Science, Engineering, and Technology, 5*(7), 19–25.
21. Hanif, S., Khedr, A. M., Al Aghbari, Z., & Agrawal, D. P. (2018). Opportunistically exploiting internet of things for wireless sensor network routing in smart cities. *Journal of Sensor and Actuator Networks, 7*(4), 46.
22. Shaikh, F. K., & Zeadally, S. (2016). Energy harvesting in wireless sensor networks: a comprehensive review. *Renewable and Sustainable Energy Reviews, 55*, 1041–1054.
23. Jing, Q., Vasilakos, A. V., Wan, J., Lu, J., & Qiu, D. (2014). Security of the internet of things: perspectives and challenges. *Wireless Networks, 20*(8), 2481–2501.
24. Lin, S., Miao, F., Zhang, J., Zhou, G., Gu, L., He, T...., & Pappas, G. J. (2016). ATPC: adaptive transmission power control for wireless sensor networks. *ACM Transactions on Sensor Networks (TOSN), 12*(1), 1–31.
25. Kalantarian, H., Motamed, B., Alshurafa, N., & Sarrafzadeh, M (2016). A wearable sensor system for medication adherence prediction. *Artificial Intelligence in Medicine, 69*, 43–52.
26. Rashid, B., & Rehmani, M. H. (2016). Applications of wireless sensor networks for urban areas: a survey. *Journal of Network and Computer Applications, 60*, 192–219.
27. Haseeb, K., Islam, N., Almogren, A., & Din, I. U. (2019). Intrusion prevention framework for secure routing in WSN-based mobile internet of things. *IEEE Access, 7*, 185496–185505.

28. Noel, A. B., Abdaoui, A., Elfouly, T., Ahmed, M. H., Badawy, A., & Shehata, M. S. (2017). Structural health monitoring using wireless sensor networks: a comprehensive survey. *IEEE Communications Surveys & Tutorials, 19*(3), 1403–1423.
29. Blikstein, P. (2013). Digital fabrication and 'making' in education: the democratization of invention. *FabLabs: Of Machines, Makers and Inventors, 4*, 1–21.
30. Razzaque, M. A., Milojevic-Jevric, M., Palade, A., & Clarke, S. (2015). Middleware for internet of things: a survey. *IEEE Internet of Things Journal, 3*(1), 70–95.
31. Hashem, I.A.T., Chang, V., Anuar, N.B., Adewole, K., Yaqoob, I., Gani, A., et al. (2016). The role of big data in smart city. *International Journal of Information Management, 36*(5), 748–758.
32. Sodhro, A.H., Luo, Z., Sangaiah, A.K., & Baik, S.W. (2019). Mobile edge computing based QoS optimization in medical healthcare applications. *International Journal of Information Management, 45*, 308–318.
33. Bera, S., Misra, S., Roy, S. K., & Obaidat, M. S. (2016). Soft-WSN: software-defined WSN management system for IoT applications. *IEEE Systems Journal, 12*(3), 2074–2081.
34. Singh, S. P., & Sharma, S. C. (2015). A survey on cluster based routing protocols in wireless sensor networks. *Procedia Computer Science, 45*, 687–695.
35. Al-Turjman, F., & Alturjman, S. (2018). Context-sensitive access in industrial internet of things (IIoT) healthcare applications. *IEEE Transactions on Industrial Informatics, 14*(6), 2736–2744.
36. Paul, P. K., Solanki, V. K., & Kumar, R. (2020). An Analytical Approach from Cloud Computing Data Intensive Environment to Internet of Things in Academic Potentialities. In *Principles of Internet of Things (IoT) Ecosystem: Insight Paradigm* (pp. 363–381). Springer, Cham.
37. Han, G., Zhou, L., Wang, H., Zhang, W., & Chan, S. (2018). A source location protection protocol based on dynamic routing in WSNs for the social internet of things. *Future Generation Computer Systems, 82*, 689–697.
38. Hodge, V. J., O'Keefe, S., Weeks, M., & Moulds, A. (2014). Wireless sensor networks for condition monitoring in the railway industry: a survey. *IEEE Transactions on Intelligent Transportation Systems, 16*(3), 1088–1106.
39. Alsheikh, M. A., Lin, S., Niyato, D., & Tan, H. P. (2014). Machine learning in wireless sensor networks: algorithms, strategies, and applications. *IEEE Communications Surveys & Tutorials, 16*(4), 1996–2018.
40. Lee, I., & Lee, K. (2015). The Internet of Things (IoT): applications, investments, and challenges for enterprises. *Business Horizons, 58*(4), 431–440.
41. Rawat, P., Singh, K. D., Chaouchi, H., & Bonnin, J. M. (2014). Wireless sensor networks: a survey on recent developments and potential synergies. *The Journal of Supercomputing, 68*(1), 1–48.
42. Turkanović, M., Brumen, B., & Hölbl, M. (2014). A novel user authentication and key agreement scheme for heterogeneous ad hoc wireless sensor networks, based on the Internet of Things notion. *Ad Hoc Networks, 20*, 96–112.
43. Li, X., Peng, J., Niu, J., Wu, F., Liao, J., & Choo, K. K. R. (2017). A robust and energy efficient authentication protocol for industrial internet of things. *IEEE Internet of Things Journal, 5*(3), 1606–1615.
44. Sobral, J. V., Rodrigues, J. J., Rabelo, R. A., Lima Filho, J. C., Sousa, N., Araujo, H. S., & Holanda Filho, R. (2018). A framework for enhancing the performance of Internet of Things applications based on RFID and WSNs. *Journal of Network and Computer Applications, 107*, 56–68.
45. Srbinovska, M., Gavrovski, C., Dimcev, V., Krkoleva, A., & Borozan, V. (2015). Environmental parameters monitoring in precision agriculture using wireless sensor networks. *Journal of Cleaner Production, 88*, 297–307.
46. Talari, S., Shafie-Khah, M., Siano, P., Loia, V., Tommasetti, A., & Catalão, J. (2017). A review of smart cities based on the internet of things concept. *Energies, 10*(4), 421.
47. Sha, K., Wei, W., Yang, T. A., Wang, Z., & Shi, W. (2018). On security challenges and open issues in Internet of Things. *Future Generation Computer Systems, 83*, 326–337.

2

Computational Intelligence in Wireless Sensor and Actuator Networks: A Thorough Review

Thakur Santosh and Sandip K. Chaurasiya

CONTENTS

DOI: 10.1201/9781003102397-2

2.1 Introduction

Wireless sensor and actuator networks (WSANs) are the natural extension of wireless sensor networks (WSNs) in which nodes are equipped with the ability to act upon the environment's current status as per the need of the intended application. Contrary to the WSNs, which comprise computationally limited devices with limited energy, WSANs include more capable nodes, resource-rich sinks, and actuators in their deployment. With the help of such sinks and actuators, the target environment can be acted upon directly to introduce the expected alteration, thus allowing a huge potential for the computation in the network.

Like WSNs, WSANs have also found their acceptance in a wide variety of applications, ranging from agriculture to autonomous animal guidance and control. Nowadays, in almost in every aspect of human lives, WSANs are proving their significance, especially in smart home applications, alarm raising applications, critical infrastructure protection, and industrial applications. However, like their predecessor, the WSANs also encounter several design challenges such as an energy-efficient network solution, scalability, coverage, and connectivity. In addition to the aforesaid design issues, WSANs suffer from reasoning and sensing limitations, limitations in performing the actuation and limitations due to network heterogeneity, etc.

Designing solutions for such problems can be perceived as optimization problems involving multiple decision variables. To solve these complex multimodal optimization problems, the new paradigm of computational intelligence (CI) has proven to be a better approach over the traditional methods. CI includes a wide variety of techniques to address the aforementioned problems such as evolutionary algorithms (EAs), swarm intelligence, biological phenomena-based metaheuristic schemes, fuzzy systems, and granular computing. Two major features are common in almost all the techniques under this umbrella of CI – inspiration from the biological events and converting imprecise solutions into approximate yet sound solutions. The simplicity and efficacy of these approaches in solving the target problems have encouraged the researchers to apply the techniques of CI very enthusiastically in the domain of WSANs.

The main idea behind the work is to throw light on the complete horizon of the CI in the domain of WSANs so that the readers might become able to explore the field and search comparatively better solutions to the existing challenges.

2.1.1 Major Contribution and Organization of the Work

The major contributions of the work are listed below:

- Discus the taxonomy of CI in detail
- Survey of the contributions made in the context of WSANs using the techniques of CI
- Identify the future research directions and opportunities

The rest of the chapter is organized into five sections. In Section 2.2 the taxonomy of CI pertaining to WSANs is discussed in detail. Thereafter, Section 2.3 discusses the major design issues in WSANs. Section 2.4 surveys the major contributions made by the researchers' community in the concerned domain. Section 2.5 lists the future research directions; and finally, Section 2.6 concludes the entire work.

2.2 Taxonomy of Computational Intelligence Techniques

CI is defined by the Institute of Electrical and Electronics Engineers (IEEE) Computational Intelligence Society as the theory, design, application, and development of biologically and linguistically motivated computational paradigms, emphasizing neural networks, connectionist systems, genetic algorithms, evolutionary programming, fuzzy systems, and hybrid intelligent systems in which these paradigms are contained under the article I and section V in its constitution [3]. As stated earlier, in almost all of the CI techniques the input is imprecise and their working is nature inspired. The aforesaid definition leads to categorization of CI techniques such as fuzzy logic, EAs, swarm intelligence, learning systems, and hybrid systems, as in Figure 2.1. In the following subsections, these methods will be discussed in detail.

2.2.1 Fuzzy Logic

Fuzzy logic refers to the technique of reasoning effectively with the uncertainty of how to deal with the fuzziness present in the system [4]. Contrary to the notion of classical crisp set wherein either the member belongs or does not belong to the set, fuzzy logic assigns a membership value (between 0 and 1) to each of the elements of the set, defining its degree of association. With the help of such an idea of degree of association, we can easily quantify the vagueness and impreciseness of the events under consideration and can deduce the decision control logic. In fuzzy logic, the assignment of a value 2 [0; 1] to the arbitrary element in the domain of discourse is facilitated by a function known as membership

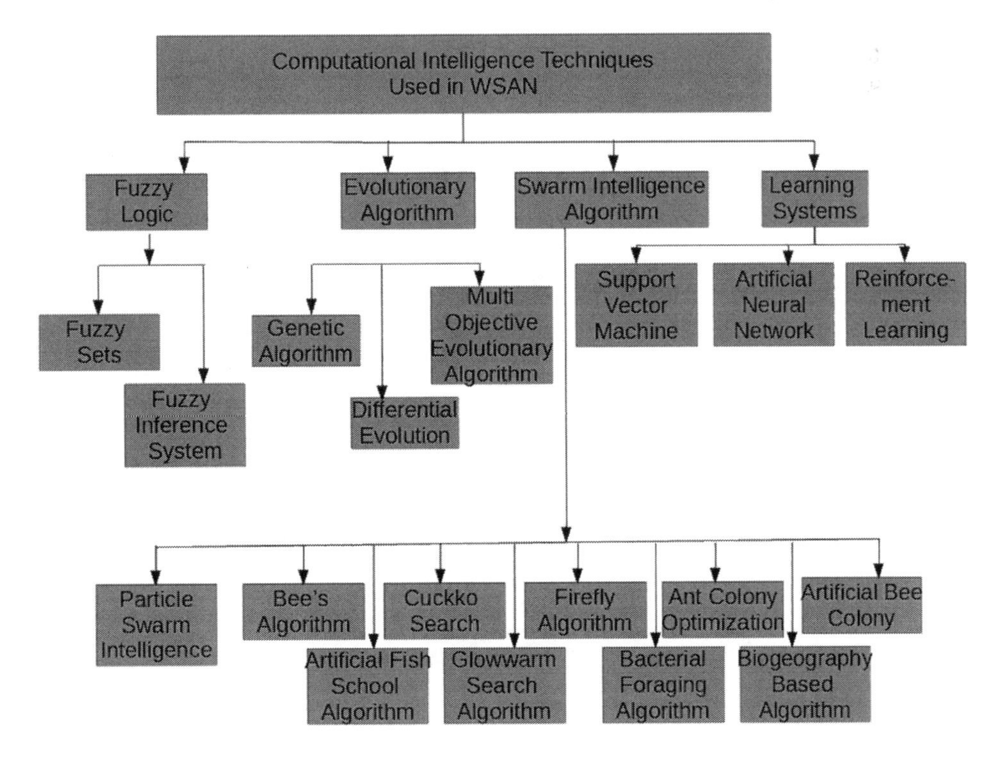

FIGURE 2.1
Categorization of CI techniques.

function. From the wide variety of available membership functions, some of the most used are impulsive membership function, triangular membership function, trapezoidal membership function, and Gaussian membership function.

Fuzzy logic plays a significant role in designing the control systems for modern applications and can be hybridized with other computational intelligence techniques. The process of converting the vector of input values into that of output values based on a set of rules can be intuited by the working of a fuzzy inference system (FIS), as in Figure 2.2:

2.2.1.1 *Fuzzy Inference System*

A FIS is used to formulate the output crisp values based on the supplied fuzzy sets along with the corresponding membership functions and a set of defined inference rules. There are majorly two popular FISs – Mamdani type [5] and Sugeno type [6]. The two types mainly differ from one another in the way how the output is produced. In a FIS, the inference rules are determined via expert knowledge or can be learned from the available data. The working of such a FIS can be illustrated as follows: At very first, via the appropriate membership functions associated with the predefined fuzzy sets, the input to the system is fuzzified. These fuzzy sets are labeled by some linguistic terms, which are further used either as the antecedents and/or consequents by the inference rules. The strength of each of the rules is computed in the sense that some of those can be proved more important in reaching a conclusion. Such computed strength of the rules is then aggregated and weighted to produce a fuzzy value. Thereafter, the obtained fuzzy value is defuzzified by employing a suitable method such as the centroid method (in Mamdani-type FIS) or weighted average of the output values from the rules (in Sugeno-type FIS).

2.2.2 Evolutionary Algorithms

EAs are inspired by the progressive method of the natural evolution of species in which the fittest can survive only. They apply almost the same procedure of natural selection to find the most appropriate solution to the concerned optimization problem. In an EA, the fitness function is the most significant concept. The fitness function is used to determine the suitability of the obtained solution. As per the problem, the maximum or the minimum value of the function can be taken as the qualifying threshold confirming the

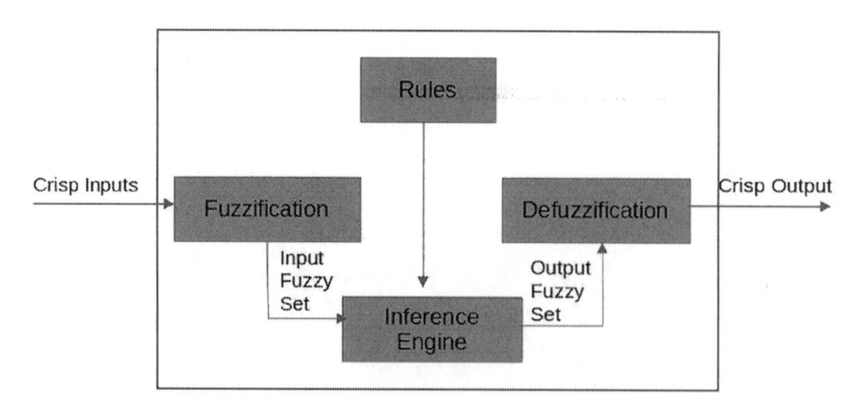

FIGURE 2.2
Fuzzy inference system – Mamdani type.

appropriateness of the solution. The working of the scheme is repeated over the iterations and stops upon encountering a predefined criterion in the sense that either the maximum number of iterations or the desired fitness values or the specified execution time is met.

At the beginning of the operation, an initial population of a definite number of individuals (also called chromosomes) is taken. Each of the individuals comprises a series of genes. The chosen population is then passed through a variation phase to proceed toward having the new individuals. The variation phase normally consists of mutation followed by crossover operations. In the mutation operation, the existing individuals are altered to obtain the newer ones; whereas, in the crossover operations, two more individuals are selected and operated accordingly to produce a new individual. The fitness of the newly generated individuals is then evaluated as per the definition of the corresponding fitness function, and based on the chosen selection policy (usually greedy selection strategy), the individuals are obtained. Thus the obtained set of new individuals defines the new generation. At the end of the operation, the individual with the best fitness value is chosen as the solution to the optimization problem. However, the generated solution is not guaranteed to be the accurate, as the chosen fitness function and the algorithm may get stuck into a local maxima/minima, not the global maxima/minima.

Some of the popular variants of this class are listed as follows.

2.2.2.1 Genetic Algorithm

Genetic algorithm is one of the first techniques in this category which is based on the principle of natural selection. They are not restricted to optimization problems only; however, they are used more popularly in this field also. They follow the traditional iterative approach of improving the existing solutions in every iteration while starting with the randomly initialized solution space. The mutation process makes small change in existing solution and the crossover phase results in the offspring solution, which may or may not be the part of the new generation depending on the fitness values.

Moreover, while applying the aforementioned approach in the concerned problem, one must take care of the bounds imposed on the values that the gene can take. In case of any violation, individuals might be adjusted or discarded accordingly.

2.2.2.2 Differential Evolution

In [7], Storn and Price proposed the differential evolution, which is evolved as one of the most popular schemes in the context of computational intelligence. The scheme starts with the initialization of basic parameters – initial random population (say, of the size N) and the number of iterations, and that of the schematic parameters-scaling factor and crossover rate. Once all the parameters are initialized, each of the solution vectors (chromosome/genome) with D decision variables, known as target vector ($\Omega_i \mid i \in 1, 2, 3,. ., N$) from the initial population (Ω), is submitted to the mutation phase. There exists a wide variety of mutation strategies based on choices made for differential vectors and their count, such as random vectors, best vectors, and target-to-best vectors. One can easily decide and utilize the most suitable strategy for his concerned problem. The target vector, once passed through the mutation phase, becomes the donor vector or the mutant vector ($\Xi_i \mid i \in 1, 2, 3,. ., N$), where the complete set of the mutant vectors is represented by Ξ. After getting through the mutation phases, the target vector is submitted for the crossover beyond which the donor vector is termed as a trial vector ($\Psi_i \mid i \in 1, 2, 3,. ., N$), and the set of such trial vectors is denoted by Ψ. The crossover strategy can be of two types – binomial (bin) and exponential (exp).

Binomial crossover

$$\psi_j = \begin{cases} \xi_j & \text{if } r \leq C_r \text{ OR } j = \delta \\ \omega_j & \text{if } r > C_r \text{ AND } j \neq \delta \end{cases} \tag{2.1}$$

where C_r is the crossover rate, δ is the randomly selected integer from the set {1, 2, 3, …, D}, r is the random number between 0 and 1, ψ_j refers to the *jth* variable of the trial vector, ξ_j refers to the *jth* variable of donor/mutant vector, and ω_j refers to the *jth* variable of the target vector.

Exponential crossover: In the exponential crossover, at very first, the *nth* variable from the donor vector is copied into the trial vector. Afterward, every subsequent variable from the donor vector is copied into the trial vector as long as the $r \leq C_r$. Once $r > C_r$, variables from the target vector are copied into the trial vector.

Depending on the chosen mutation and crossover strategies, many variants for the differential evolution have been proposed. To discriminate among those variants, Kendall's like standard notation, *DE/x/y/z*, is used. Here, *DE* indicates the differential evolution, *x* refers to the selected mutation strategy, *y* denotes the number of differential vectors being used in the mutation scheme, and *z* refers to the adopted crossover strategy. A few major variants being popularly used are as follows: *DE/rand/1/bin*, *DE/best/1/bin*, *DE/best/2/exp*, and *DE/target – to – best/1/bin*, etc.

2.2.2.3 Multi-Objective Evolutionary Algorithms

Most of the modern real-world optimization problems consist of satisfying multiple objectives simultaneously, which might be conflicting. For such optimization problems, instead of having a single optimal solution, it is better to have a set of alternative solutions. This set of alternative solutions, popularly known as "Pareto-optimal" solutions, is optimized in the sense that there won't be any solution considering all the objectives in the search space, which can be considered more superior.

Mathematically the notion of a multi-objective optimization problem can be defined as follows:

$$\text{maximize / minimize } y = f(x) = (f_1(x), f_2(x), f_3(x), \ldots, f_n(x)) \tag{2.2}$$

such that y comprises n objective functions $f_i : \Xi \rightarrow \Omega$ x, and y.

Here, $x \in \Xi$ is called the decision vector, $y \in \Omega$ is termed as the objective vector, Ξ refers to the parameter or decision space, and Ω is known as objective space.

The objectives for (2.2) might often conflict with each other in such a way that the improvement of one objective may result in the degradation of another. Thus, a single solution, optimizing all the objectives simultaneously, might not exist. Instead, the best trade-off solutions as explained above are called the Pareto-optimal solutions and are of much importance to the decision-maker. To address this class of problem, an extension of the traditional EA, the multi-objective evolutionary algorithm (MOEA) has been adopted by the research community. Some methods can transform the multi-objective optimization problems to a single-objective optimization problem via specialized aggregation functions such as a weighted sum approach and target vector optimization.

2.2.3 Swarm Intelligence

Similar to EAs, swarm intelligence is also a nature-inspired technique. Contrary to the EA in which the phenomenon of biological evolution is limited, swarm intelligence is motivated by the collective behavior of the social animal societies. However, the working of both the schemes resembles closely in the sense that they both start with an initial population of finite size and follow the method of iterative progression toward the most optimal solution. In swarm intelligence, fitness function becomes the most significant concept as it guides the entire search process to the final solution as in the approach of the EA. Some of the popular swarm intelligence algorithms are given in next subsections.

2.2.3.1 Particle Swarm Optimization

In [8], Kennedy and Eberhart presented the particle swarm optimization (PSO), which is a guided random search algorithm inspired by the collective behavior of the social animal societies. However, it usually models the social behavior of fish schooling or birds flocking. In this technique, every particle is characterized by a position and a velocity associated with it. The particles may change their respective position by tuning their velocity. The particles are also equipped with the capability to remember the best position achieved by it and can communicate the best position explored by them to their peer ones. Like the differential evolution technique, the initial position and velocity population vectors (say, Ω and Θ, respectively) are generated randomly but within the available search space. After such random initialization of preliminary populations, the fitness of each position vector is computed. Here, it is to be noted that if the initial population referring to the initial positions of the particle is of the size N, each with D decision variables then Ω would be a $[N \times D]$ matrix, whereas the fitness matrix (Γ) referring to the fitness values of position vectors would be a $[N \times 1]$ matrix. Along with the maintenance of the Ω, Θ, and Γ, two temporary matrices – Ω_{best} and Γ_{best} are also maintained in the PSO for the determination of the final solution vector. Here, Ω_{best}, initially being the same as that of Ω, keeps track of the best position explored by the vectors themselves; hence known as personal best, and Γ_{best} maintains the corresponding values attained by the vectors in Ω_{best}. From the Ω_{best}, a vector is chosen with the best possible fitness value, which is termed as the **global best** and denoted by Λ_{best}.

The scheme proceeds in a way that in every round, each velocity vector say Θ_i of Θ is processed as per the following equation:

$$\Theta_i = \omega.\Theta_i + c_1.r_1.(\Omega_{best,i} - \Omega_i) + c_2.r_2.(\Lambda_{best} - \Omega_i) \tag{2.3}$$

where $\omega \in (0, 1)$ is the particle inertia, c_1 and c_2 are the acceleration coefficients $\in (0, 2)$, and r_1, r_2 are random row vectors of the length D. Once Θ_i is obtained as per (2.3), the corresponding position vector say Ω_i from Ω is also updated as follows:

$$\Omega_i = \Omega_i + \Theta_i \tag{2.4}$$

After obtaining Ω_i, the population vector is updated irrespective of its current fitness value (i.e., no greedy approach is followed for updating the Ω); however, Ω_{best}, Γ_{best}, and Λ_{best} are updated following the greedy approach as follows:

$$\Omega_{best,i} = \Omega_i \text{ if } ff^n(\Omega_i) < ff^n(\Omega_{best,i}) \tag{2.5}$$

And,

$$\Lambda_{best} = \Omega_{best,i} \text{ if } ff^{n}\left(\Omega_{best,i}\right) < ff^{n}\left(\Lambda_{best}\right) \tag{2.6}$$

After executing the aforesaid steps for a definite number of rounds, one may have the final solution.

2.2.3.2 Ant Colony Optimization

Ant colony optimization (ACO) is another variant of the class, swarm intelligence techniques. ACO imitates the foraging behavior of real ants and is a multi-agent method. In the search for their food, ants travel across the field and deposit on a ground a chemical substance, called pheromone. Upon encountering the food, they start traveling back to their place while depositing the rational amount of pheromone justifying the quality of food. ACO is a probabilistic model and iteratively assigns higher pheromone values to a good solution. These higher pheromone concentrations help the algorithm converge to promising regions of the underlying optimization graph. Several versions of the ACO have been created by the researchers by varying the pheromone model.

2.2.3.3 Artificial Bee Colony

The artificial bee colony (ABC) algorithm is another swarm intelligence-based metaheuristic search technique to solve the optimization problems. The technique is inspired by the intelligent foraging behavior of honey bees by which the bees are divided into three major categories – employed bees, onlooker bees, and scout bees. The employed bees confirm the source of food by dancing into a specified area that is consistently being monitored by the onlooker bees. The onlooker bees exploit the food position based on the information received by the employed bees. In case of abandonment of existing food sources, scout bees start exploring new sources. It is believed that the number of employed bees is equal to food position and one employed bee is mapped to a single food source. To apply the ABC algorithm in the context of the optimization problem, the optimization problem is first converted to the problem of finding the best parameter vector minimizing the objective function. Here, the objective being represented by fitness value refers to the quality of a food source. The artificial bees randomly discover a population of initial solution vectors and iteratively upgrade them via the strategies – moving toward better solutions by employing a neighbor search mechanism while abandoning the poor solutions.

2.2.4 Learning System or Machine Learning

In the learning system (LS) or machine learning (ML), the systems learn from the available set of data and act accordingly. ML refers to the process of unveiling the relationship between the numerical features and the states or objects in the way the living organism does. However, the relationship is not known in advance but has to be discovered or learnt from the set of available data. ML is classified into three categories, namely, supervised, semi-supervised, and unsupervised.

In the supervised model, the machines are trained with class labels, and in the semi-supervised model, the same is achieved with the portions of the label and unlabeled data. In the unsupervised model, the machines are trained with the unlabeled data.

Two types of learning system that are popularly being used in the domain of interest are WSANs.

2.2.4.1 Artificial Neural Networks

Artificial neural networks (ANNs) are primarily inspired by how the brain functions in the human body. Inside the brain of the human body, there are neurons, made up of cells, which can process the brain signals into the actionable messages via regulating the human body. The traditional neural network is composed of interconnected neurons, which are classified into three major layers – an input layer, an output layer, and one or more hidden layers, as in Figure 2.3.

Similarly, ANNs are composed of hundreds or thousands of artificial neurons (termed as processing units) interconnected by the nodes. These processing units are made up of input and output units. Based on an internal weighting system, the input units receive information in various forms and structures. The neural network learns about the information presented to generate an output report. ANNs use a set of learning rules called backpropagation to perfect their output results, similar to the way humans need rules and guidelines to come up with a result.

2.2.4.2 Reinforcement Learning

Reinforcement learning (RL) refers to how the intelligent agents take the actions in an environment while maximizing the notion of cumulative reward. The intelligent agent perceives from the environment and optimally acts upon the environment to receive the maximum possible reward.

RL algorithms are significantly suitable in multi-agent systems where the individual nodes do not communicate or the social learning of all the individual nodes can be integrated, enabling these agents to perform better.

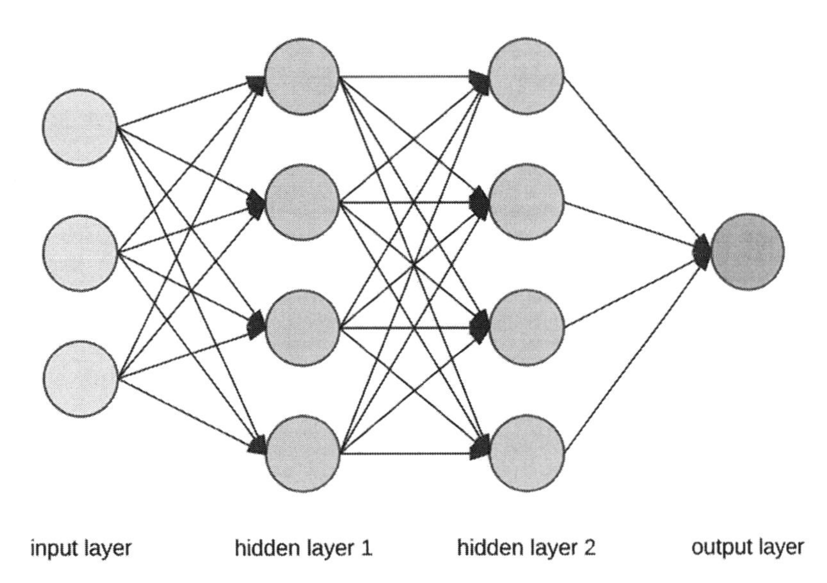

FIGURE 2.3
An artificial neural network.

2.2.5 Hybrid System

To compensate for the shortcomings of an individual computational intelligence technique, the concept of a hybrid system (HS) is gaining huge popularity among researchers. The HS involves the integration of two or more different computational intelligence techniques like evolutionary algorithms combined with swarm intelligence algorithms (due to the similar nature of work); fuzzy logic with the neural networks to solve the complex problems, etc.

The significance of the hybrid systems is evident from the huge increase in the deployment of hybrid approaches to solve the optimization problems of WSAN scenarios.

2.3 Design Issues in Wireless Sensor and Actuation Networks

2.3.1 Wireless Sensor and Actuation Networks

The main idea behind the deployment of wireless sensor networks is to monitor the surroundings and forward the collected information to a centralized base station for further processing and user access. However, it might not be always sufficient just to sense and observe the target area as in the routine applications of WSNs, but to act upon the surroundings as per the collected information would be required by the intended application. For example, in a fire alarm and protection system, it might be required to turn on the water sprinklers based on the fire confirmation by the deployed sensors. To perform actions based on the sensor data, actuators are used and such an extension of traditional WSN becomes a wireless sensor and actuator network. The WSAN enables the application to sense, interact, and act upon the target environment.

The WSAN is usually a heterogeneous network of the sensor nodes and the actuators deployed along with one or more sink nodes, as in Figure 2.4. Both the sensor nodes and the actuators in WSANs are equipped with certain processing and data communication abilities. Normal sensor nodes collect the surrounding data and forward the data to the actuators via a single-hop or multi-hop. Thereafter, the actuators decide upon how to alter the current state of the system based on the received data and built-in logic. The overall responsibility of the base station/sink is to monitor and manage the network operations thoroughly.

In traditional WSANs, the deployed sensor nodes remain static, whereas the actuators may be mobile. Also, the actuators are enriched with additional resources to perform computationally rich operations and are more autonomous. Moreover, depending upon the application of interest and nodes' ability, the nodes can play any of the roles mentioned earlier (i.e., of the normal sensor node, of the actuator, and that of the sink as well).

2.3.2 Design Issues

In addition to the challenges of WSNs, WSANs suffer from more design issues, as listed in Figure 2.5. It can be easily intuited that due to the additional movement of actuators, management of network topology, and hence the communication within the network becomes more challenging in WSANs with respect to that in WSNs. Similarly, other challenges can be broadly categorized into five major classes – actuation, communication, localization, mobility, and topology control.

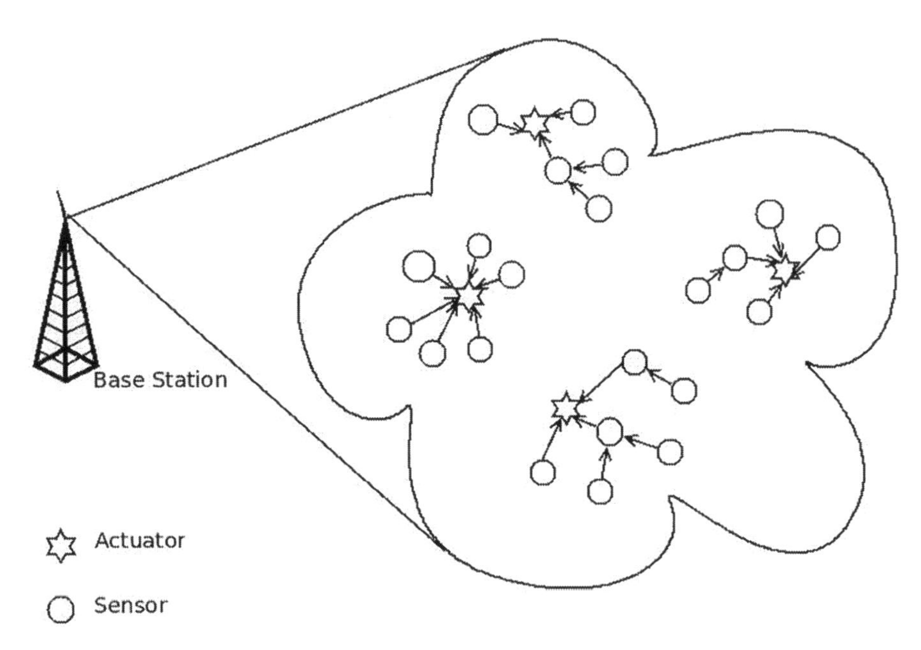

FIGURE 2.4
Wireless sensor and actuator network [10].

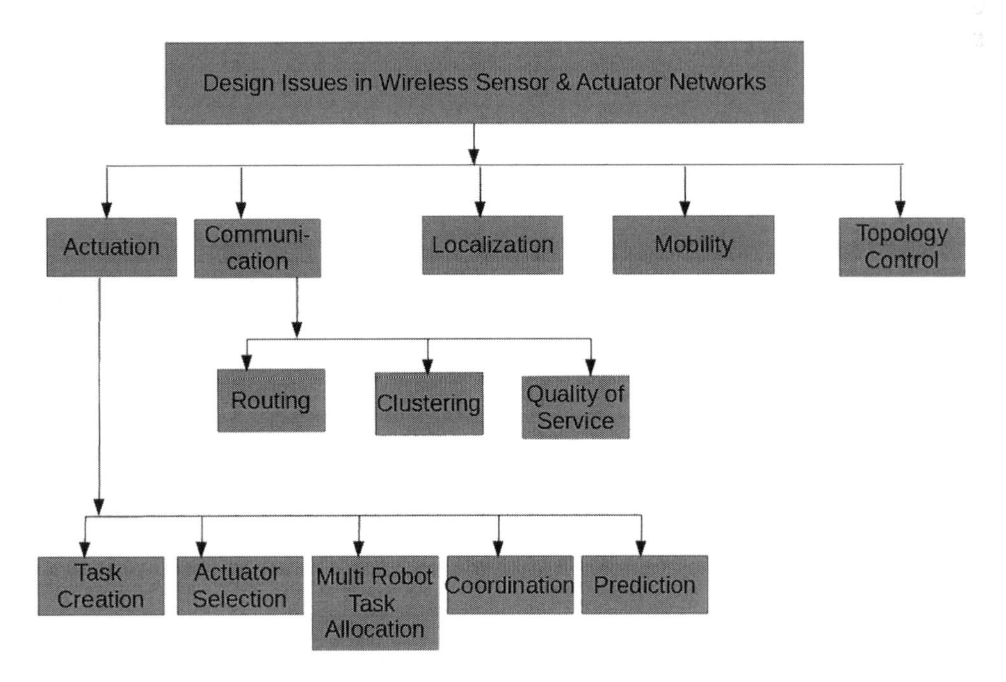

FIGURE 2.5
Design issues in wireless sensor and actuation networks.

FIGURE 2.6
Actuation in WSANs.

2.3.2.1 Actuation

The process of actuation is the main discriminator between WSNs and WSANs. The process starts with the data received from the sensor nodes and reaches the sink node from where instructions are provided to the actuator to change the system state. The entire process of actuation is summarized in Figure 2.6.

a. **Task creation:** A task refers to an action that must be executed by the actuator(s) based on the readings received from the sensors in the field. There may be two types of tasks being generated in the field – proactive tasks and reactive tasks. Proactive tasks indicate that some events are about to happen (anticipated based on sensors' readings) and actuators have to generate a task to counter the events. Contrary to this, reactive tasks refer to the tasks that have to be generated to counter the effect of the events, which have already been happened in the field. This step of task creation requires data fusion. Moreover, the task can be altered or modified at any time based on the updated information by the sensor nodes; and hence, one must consider the unreliability of the sensor data while creating the tasks.
b. **Actuator selection:** Immediately after the creation of tasks, there comes the problem of selecting a suitable set of actuators in a way to effectively execute the tasks. To effectively contribute toward the intended application, the interaction among the set of chosen actuators and that with the remaining ones in the network must be considered accordingly. The aforementioned requirement is termed a combinatorial optimization problem. Moreover, the issue of inaccuracy and unavailability of the received information by the sensors should also be brought into consideration in this phase of actuator selection.
c. **Task allocation:** Task allocation refers to the process of assigning one or more tasks to the chosen set of actuators in an optimized and timely manner.
d. **Actuator coordination:** Actuator coordination indicates the requirement of coordinating with one another within the set of chosen actuators while taking support from the network in order to get the tasks accomplished. This is a highly significant step required throughout the task resolution to avoid conflicts.
e. **Event prediction:** Event prediction is a central requirement of the proactive systems, which ensures the timely prediction of events influencing the rest of the WSANs.

2.3.2.2 Communication

In WSNs, the nodes are energy-constrained and computationally limited. This makes the routing of data to the sink node a challenging task. Since WSANs are the extension of WSNs in the sense that besides having sensor nodes and sink, they also contain the specialized actuator nodes. In many of their applications, the actuators might be mobile, creating another level of complexity in designing communication protocols for the networks. The major problems exacerbating communication in the WSANs are listed as follows:

a. **Routing:** Like in WSNs, routing is also a major problem in WSANs. Being an extension of WSNs, WSANs find some of the routing issues similar to those in WSNs, such as energy holes. In the multi-hop approach of transferring data to the sink, the nodes near the sink are frequently used as the relay nodes. This results in early depletion of nodes' battery disconnecting the sink from the rest of the network. Thus, the problem of energy holes might be taken into account while developing routing solutions in the WSANs. In addition to this, since the nodes can take on any role – normal node, actuator, and sink in the network – there might be scenes of multiple sinks equipped with mobility, which should also be taken care of adequately.
b. **Clustering:** Clustering of the nodes has been established as a very popular approach for having energy-efficient network operations in WSNs. The same can also be applied to the context of WSANs. However, the heterogeneity and the mobility present in the network should be treated appropriately.
c. **Quality of service:** Like other networks, WSANs also expect quality metrics such as high throughput, low error rates, low latency, and low jitter to measure the performance. For example, actuators respond to the happenings in the field, which might be a requirement with the least possible delay.

2.3.2.3 Localization

Localization is a major issue in WSNs and thus in WSANs. Being equipped with the nodes' current location is always desirable and may be of great benefit in many of the applications. However, acquiring nodes' location requires installing Global Positioning System (GPS) devices or some reference frame. GPS might result in a quick drain of the nodes' battery leading to the early death of nodes (i.e., the reduced network lifetime). Moreover, GPS may result in inaccurate information, especially for the indoor and/or urban applications. Hence the problem of localization has to be tackled in a smarter way, ensuring a good trade-off among parameters such as accuracy, cost, and energy efficiency.

2.3.2.4 Mobility

In WSANs, actuators and sink(s) might be mobile in the sense that they can travel across the sensing field. Such movement can create challenges of formulating a trajectory to ensure efficient data collection and prompt actuation while minimizing the energy consumption in the network.

2.3.2.5 Topology Control

Topology control is a major issue in WSANs because WSANs are composed of heterogeneous nodes such as static sensor nodes, static/mobile actuators, and static/mobile sinks. Although the mobile actuators and sinks can result in improved coverage, the same results in the increased complexity of topology management.

2.4 CI Contributions in WSANs

As discussed in Section 2.3, the main design challenges into the WSANs can be classified in five major categories – actuation, communication, localization, mobility, and topology control. Since the inception of WSANs, a huge number of computational intelligence-based contributions have been recorded in the literature. In this section, the major contributions addressing these issues will be discussed in the aforementioned order.

2.4.1 Actuation

In [11], McCausland et al. proposed an auction-based scheme to select the most eligible actuators from the set of available actuators. The scheme is inspired by the auction-bidding process in which an auction is initiated after the task is created. Immediately after the task creation, all the actuators compute their respective bid value based on the parameters – current residual energy, distance from the event, and redundant coverage via a FIS. The actuator with the highest bid value is selected and allocated the tasks. Here, for the task allocation process, non-dominated sorting genetic algorithm-II is used while pursuing the problem as a multi-objective optimization one.

In [12], Wang and Wang proposed an intelligent residential lighting control system based on a ZigBee wireless sensor network and a fuzzy controller. In their proposal, more emphasis is on the task creation. The data collected by the deployed sensor nodes are forwarded to the FIS controller, which creates the assignment for the specific actuator node. However, the proposed work lacks any supporting simulation.

In [13], La et al. proposed a target tracking application of the WSANs. The deployed sensor nodes track the current location information of the target. Based on this collected information, the next expected target location is computed. Thereafter, the genetic algorithm is used to determine the most suitable mobile node to be relocated to cover the expected path of the target. Both actuator selection and task allocation are considered simultaneously by the genetic algorithm to optimize the distance to the target object and the time to catch up with the target.

In [14], Venayagamoorthy et al. proposed a control system for the WSAN, which utilizes a combination of fuzzy logic and swarm intelligence techniques for the collective robotic search application. The proposed work calls two methods: the first method employs a particle swarm optimization-based engine augmented with fuzzy logic and the second one is based on a FIS. The first method explores the search space as per the logic of fuzzy PSO; meanwhile, in the second method, mobile nodes compute their fitness, and the sensor nodes are attracted to the node with the highest fitness value.

2.4.2 Communication

In this subsection, major contributions addressing the issue of communication via three different aspects – clustering, routing, and quality of service are discussed as follows:

In [15], Obaidy and Ayesh proposed a metaheuristic clustering algorithm based on the EA and particle swarm optimization (PSO) for the mobile WSANs. The proposed method generates the clusters using the EA and then coordinates the movement of clusters via PSO. For the formation of clusters, a genetic algorithm (GA) has been used particularly. In the respective fitness function of GA, a number of cluster heads along with their distances from the member nodes are considered as the parameters to produce effective clusters.

Efficacy of the proposed methodology in the formulation of clusters has been established through an extensive set of experimentation.

In [16], Srivastava and Sudarshan proposed a genetic fuzzy system-based routing protocol for mobile WSANs, emphasizing clustering to achieve energy efficiency. In their work, a two-phase method is proposed for determining suitable clusters. In the first phase, a FIS is employed by each of the nodes to determine its suitability to be a cluster head based on its residual energy, local node density, distance to other nodes, and mobility. Then, a genetic algorithm is employed for the final selection of cluster heads; in this, the candidature of every solution is evaluated in terms of the number of cluster heads, mean communication energy required, and speed of the cluster heads.

In [17], Yang et al. proposed a swarm intelligence-based routing protocol for the networks with mobile sinks. In their proposal, they modified the PSO technique to ensure a path with higher residual energy. Moreover, by varying the speed of mobile sinks and message bandwidths, they conclude that the message bandwidths affect the network performance more in respect of mobile sinks' speed. Also, the supremacy of the proposed scheme is established over the state-of-art schemes through an extensive set of experimentation.

In [18], Xia et al. proposed a QoS aware communication protocol for the WSANs. The proposed scheme manages the impact of uncertain changes in the traffic load on QoS of WSANs via fuzzy logic. It utilizes a fuzzy logic controller inside each of the source sensor nodes to adapt the sampling period to the deadline miss ratio associated with data transmission from the sensor to the actuator. The objective is to provide reliable QoS for the communication between the sensor and the actuator ensuring timely task allocations and situational awareness.

2.4.3 Localization

The major CI contributions toward the localization of nodes in WSANs are as follows:

In [19], Herrero and Martínez dealt with the localization as a problem of fuzzy estimation in which the node's position and measurements were described by the fuzzy sets. Here, fuzzy logic has been adopted to deal with the uncertainty present in the information. Afterward, fuzzy densities are applied over all the possible locations of a node, which are treated as the vertices of a Voronoi tessellated environment. Now the fuzzy densities are varied upon receiving the packets from the anchor nodes (the nodes with the known locations) by using received signal strength.

The proposed method is used for the mobile nodes' localization in WSANs, especially in indoor spaces. The proposed scheme has been established over the popular Monte Carlo localization approach through real-time testing with the mobile nodes.

In [20], Karedla and Anuradha proposed a two-step method for the localization of mobile nodes. In the first step, the weighted centroid algorithm is used for the computation of the position estimate of the nodes. Then, a genetic algorithm is used for the refinement of position estimate. In the applied genetic algorithm, a fitness function is designed for the estimate refinement based on the difference between the estimated locations and the received signal strength indication (RSSI) inferred distance to the anchor nodes.

Through an extensive set of simulations, the supremacy of the scheme has been proved over the state-of-art schemes.

In [21], Kulkarni and Venayagamoorthy proposed a nature-inspired iterative algorithm for the autonomous deployment and localization of sensor nodes. In his proposed scheme, they consider a network composed of three types of nodes – normal sensor nodes, anchor nodes, and an unmanned aerial vehicle (UAV) that deploys the nodes. As per the scheme, as soon as a node falls into the communication range of three or more nodes, it will initiate the process of its localization by creating a position estimate based on the distances from them. Thereafter, either the bacterial foraging algorithm (BFA) or the PSO is used to further correct the estimation.

With the deployment of new nodes in the network, previously localized nodes act like the anchor nodes helping out the new nodes in determining their respective location.

In [22], Qi et al. treated the problem of nodes' localization as a traveling salesman problem (TSP). The idea behind the scheme is quite simple; wherein the mobile anchor nodes visit each of the deployed nodes requiring localization by passing within a threshold distance of it. Thereafter, the visiting anchor nodes localize the nodes as per their known positions. To solve the aforesaid TSP problem, a hybrid of an ABC and the genetic algorithm has been applied.

2.4.4 Mobility

This subsection describes the CI solutions mainly addressing the issue of sink mobility in WSANs.

In [23], Alnuaimi et al. proposed a bio-inspired algorithm to tackle the problem of sink mobility in the sensor networks. The scheme starts with predefined clusters that occur when the network has already been partitioned into an appropriate number of clusters with designated cluster heads. The problem of sink mobility is then perceived as a TSP problem, with the cluster heads being treated as the waypoints for which a genetic algorithm is applied as the solver. The proposed scheme successfully optimizes the time needed to complete the trajectory and the energy usage of the cluster heads.

In [24], Comarela et al. considered the concerned problem of sink mobility in sparse networks that lack an adequate number of nodes for the cluster formation or the establishment of reliable communication paths. The problem is then formulated as the TSP-neighborhood problem, and ACO has been applied to solve the same while minimizing the length of the sink's trajectory.

In [25], Abo-Zahhad et al. proposed an intelligent system inspired by the artificial immune system (AIS) algorithm for the parallel optimization of sink path along with the network and the number of cluster heads. The proposed solution is then evaluated on the required communication energy as per the first-order radio model [26–28] and the number of control messages. The performance of the proposed scheme is compared to that of the schemes such as [29–31] and the supremacy of the proposed scheme is verified.

2.4.5 Topology Control

The major CI contributions addressing the topology control are as follows:

In [32], Ray and De proposed a swarm intelligence-based scheme for the network reconfiguration with mobile nodes for the enhanced and optimized coverage of the target area. They modified the glowworm swarm optimization (GSO) technique to achieve their intended objective of optimizing the distance traveled by the mobile nodes, the energy consumption for the network reconfiguration, and the redundant coverage of the solution. An extensive set of simulations has been performed with varying node densities in the network to prove the supremacy of the proposed method over the existing methods.

In [33], Ni et al. introduced another swarm intelligence-based method to further optimize the coverage in the network. In their proposed method, PSO is modified to the coverage optimization with mobile nodes. The PSO figures out the suitable nodes' position for the optimal coverage and minimal moving distance. The performance of the proposed method has been established as superior to the plain PSO technique and another PSO-based coverage enhancement method through an extensive set of experiments.

In [34], Katsuma et al. put forth a genetic algorithm-based algorithm with an objective of lifetime enhancement of the network with mobile nodes. In their scheme, the main emphasis is on optimizing the fitness function by taking into account two main parameters – energy consumption and network coverage. Via providing a connected tree comprised of nodes to the sink and k-coverage over the target AOI, the proposed scheme intelligently addresses the communication problem. The working of the proposed scheme has also been validated experimentally.

In [35], Singh and Kumar address the problem of topology control by devising a node deployment algorithm via a UAV. The scheme initiates by taking the pictures of the target AOI, which are segmented afterward accordingly. A hybrid of GA and PSO is applied to figure out the optimal positions for the improved coverage from the segmented image. A UAV is called upon then for the deployment of nodes in those identified locations in the network.

2.5 Future Research Directions

From the discussion presented in the previous sections, the future research trends can be identified as follows:

- The process of actuation starts with the task creation, which is comparatively a more ambiguous and uncertain step tightly coupled with the data fusion and event prediction. To handle this ambiguity and uncertainty, a learning system and fuzzy logic can be applied to a greater extent.
- Task creation, actuator selection, and task allocation can be jointly addressed by using EAs.
- FIS and/or ANN can be exploited for the design of control systems for the actuators that allow them to bid for the tasks created.
- EAs and/or swarm intelligence algorithms can be applied to solve routing problems as it might be required to evaluate multiple candidate routes to figure out the most appropriate one.
- EAs and/or the swarm intelligence techniques along with suitable fitness functions can be applied for the clustering of the nodes that is perceived as a multimodal optimization problem.
- Fuzzy logic and ANNs can be exploited to deal with the inherent unreliable communication in WSANs, where information accuracy and data availability must be taken into account.
- Fuzzy logic and/or learning systems can be useful in determining the nodes' position directly (i.e., without pursuing any prior estimate as it requires dealing with uncertainty).
- EAs and/or swarm intelligence-based techniques are more useful in determining the nodes' location when prior estimates are created and those are refined afterward for more accurate solutions.

- The sink mobility can be pursued like the TSP since out of multiple available routes, the sink has to choose the best one visiting all the static nodes. Thus, EAs and/or swarm intelligence techniques can be further explored to address the aforesaid as TSP is a combinatorial optimization problem.
- EAs and/or swarm intelligence-based techniques can be used also in the context of topology control as the problem can easily be mapped as a TSP variant.

2.6 Conclusion

In this work, several applications of computational intelligence techniques have been reviewed pertaining to WSANs. The survey has been presented classifying the issues/challenges of the WSANs in five major categories – actuation, communication, localization, mobility, and topology control. To deal with the aforesaid challenges, various CI techniques such as fuzzy logic, EAs, swarm intelligence, learning systems, and hybrid systems have been investigated through a wide variety of contributions already made in this domain. From the study of available literature, it can be concluded that the various CI techniques, especially the EAs and swarm intelligence techniques, are very effective in addressing the problems of WSANs. Moreover, the challenges and the open research issues along with the future research directions have also been listed extensively, encouraging the readers to explore and discover the new heights in the domain.

References

1. I. F. Akyildiz, W. Su, Y. Sankarasubramaniam, and E. Cayirci, A Survey on Sensor Networks, IEEE Communication Magazine, vol. 40, no. 8, pp. 393–422, 2002, https://doi.org/10.1109/MCOM.2002.1024422
2. A. Dumka, S. K. Chaurasiya, A. Biswas, and H. L. Mandoria, A Complete Guide to Wireless Sensor Networks: From Inception to Current Trends, 1st Edition; CRC Press, Boca Raton, FL, 2019.
3. J. Kacprzyk, and W. Pedrycz, Springer Handbook of Computational Intelligence. Heidelberg, Germany: Springer, 2015.
4. L. A. Zadeh, Fuzzy logic = computing with words, IEEE Transactions on Fuzzy Systems, vol. 4, no. 2, pp. 103–111, May 1996.
5. E. H. Mamdani, Application of fuzzy logic to approximate reasoning using linguistic synthesis, IEEE Transactions on Computers, vol. C-26, pp. 1182–1191, 1977.
6. T. Takagi, and M. Sugeno, Fuzzy identification of systems and its applications to modeling and control, IEEE Transactions on System, Man, and Cybernetics, vol. SMC-15, no. 1, pp. 116–132, Jan./Feb. 1985.
7. R. Storn, and K. Price, Differential evolution – a simple and efficient heuristic for global optimization over continuous spaces, Journal of Global Optimization, vol. 11, no. 4, pp. 341–359, 1997.
8. J. Kennedy, and R. Eberhart, Particle Swarm Optimization, in Proc. of ICNN'95- Int. Conf. on Neural Networks, Australia, vol. 4, 1995, pp. 1942–1948.
9. D. Karaboga, and B. Basturk, A powerful and efficient algorithm for numerical function optimization: Artificial bee colony (ABC) algorithm, Journal of Global Optimization, vol. 39, no. 3, pp. 459–471, 2007.

10. X. Feng. QoS challenges and opportunities in wireless sensor/actuator networks, Sensors, vol. 8, pp. 1099–1110, 2008. https://doi.org/10.3390/s8021099
11. J. McCausland, R. Abielmona, R. Falcon, A. M. Cretu, and E. Petriu, Auction-based node selection of optimal and concurrent responses for a risk-aware robotic sensor network, in Proc. IEEE Int. Symp. Robot. Sensors Environ., Washington, DC, USA, 2013, pp. 136–141.
12. Y. Wang and Z. Wang, Design of intelligent residential lighting control system based on ZigBee wireless sensor network and fuzzy controller, in Proc. Int. Conf. Mach. Vis. HumanMach. Interface (MVHI), Kaifeng, China, 2010, pp. 561–564.
13. H. M. La, T. H. Nguyen, C. H. Nguyen, and H. N. Nguyen, Optimal flocking control for a mobile sensor network based a moving target tracking, in Proc. IEEE Int. Conf. Syst. Man Cybern. (SMC), San Antonio, TX, USA, 2009, pp. 4801–4806.
14. G. K. Venayagamoorthy, L. L. Grant, and S. Doctor, Collective robotic search using hybrid techniques: Fuzzy logic and swarm intelligence inspired by nature, Engineering Applications of Artificial Intelligence, vol. 22, no. 3, pp. 431–441, 2009.
15. M. A. Obaidy, and A. Ayesh, Energy efficient algorithm for swarmed sensors networks, Sustainable Computing: Informatics and Systems, vol. 5, pp. 54–63, Mar. 2015.
16. J. R. Srivastavam, and T. S. B. Sudarshan, A genetic fuzzy system based optimized zone based energy efficient routing protocol for mobile sensor networks (OZEEP), Applied Soft Computing, vol. 37, pp. 863–886, Dec. 2015.
17. H. Yang, F. Ye, and B. Sikdar, A swarm-intelligence-based protocol for data acquisition in networks with mobile sinks, IEEE Transactions on Mobile Computing, vol. 7, no. 8, pp. 931–945, Aug. 2008.
18. F. Xia, W. Zhao, Y. Sun, and Y. C. Tian, Fuzzy logic control based QoS management in wireless sensor/actuator networks, Sensors, vol. 7, no. 12, pp. 3179–3191, 2007.
19. D. Herrero, and H. Martínez, Range-only fuzzy Voronoi-enhanced localization of mobile robots in wireless sensor networks, Robotica, vol. 30, no. 7, pp. 1063–1077, 2011.
20. S. R. Karedla, and S. Anuradha, Localization error minimization using GA in MWSN, International Journal of Advanced Electrical and Electronics Engineering, vol. 2, no. 4, pp. 17–24, 2013.
21. R. V. Kulkarni, and G. K. Venayagamoorthy, Bio-inspired algorithms for autonomous deployment and localization of sensor nodes, IEEE Transactions on Systems, Man, and Cybernetics, Part C (Applications and Reviews), vol. 40, no. 6, pp. 663–675, Nov. 2010.
22. R. Qi, S. Li, T. Ma, and F. Qian, Localization with a mobile anchor using ABC-GA hybrid algorithm in wireless sensor networks, Journal of Advanced Computational Intelligence and Intelligent Informatics, vol. 16, no. 6, pp. 741–747, 2012.
23. M. Alnuaimi, K. Shuaib, K. Alnuaimi, and M. Abdel-Hafez, Data gathering in delay tolerant wireless sensor networks using a ferry, Sensors, vol. 15, no. 10, pp. 25809–25830, 2015.
24. G. Comarela, K. Gonalves, G. L. Pappa, J. Almeida, and V. Almeida, Robot routing in sparse wireless sensor networks with continuous ant colony optimization, in Proc. Genet. Evol. Comput. Conf., 2011, pp. 599–605.
25. M. Abo-Zahhad, S. M. Ahmed, N. Sabor, and S. Sasaki, Mobile sink based adaptive immune energy-efficient clustering protocol for improving the lifetime and stability period of wireless sensor networks, IEEE Sensors Journal, vol. 15, no. 8, pp. 4576–4586, Aug. 2015.
26. S. K. Chaurasiya, T. Pal, and S. D. Bit, An Enhanced Energy-Efficient Protocol with Static Clustering for WSN, in Proceedings IEEE Xplore, Intl Conf. of Information Networking (ICOIN), Kuala Lumpur, Malaysia, March 2011, pp. 58–63, https://doi.org/10.1109/ICOIN.2011.5723134
27. S. K. Chaurasiya, J. Sen, S. Chatterjee, and S. D. Bit, EBLEC: An Energy-Balanced Lifetime Enhancing Clustering for WSN, in Proceeding IEEE Xplore, 14th Intl Conf. on Advanced Communication & Technology-2012, PyeongChang, Korea (South), Feb. 2012, pp. 189–194, INSPEC Accession Number: 12656578.
28. S. K. Chaurasiya, A. Biswas, and P. K. Bandopadhyay, International Conference on VLSI, Microwave and Wireless Technologies 2021 (ICVMWT-2021), March 20–21, 2021.

29. W. B. Heinzelman, A. P. Chandrakasan, and H. Balakrishnan, An application-specific protocol architecture for wireless microsensor networks, IEEE Transactions on Wireless Communications, vol. 1, no. 4, pp. 660–670, Oct. 2002
30. J. L. Liu, and C. V. Ravishankar, LEACH-GA: Genetic algorithm-based energy-efficient adaptive clustering protocol for wireless sensor networks, International Journal of Machine Learning and Computing, vol. 1, no. 1, pp. 79–85, Apr. 2011.
31. K. G. Vijayvargiya, and V. Shrivastava, An amend implementation on LEACH protocol based on energy hierarchy, International Journal of Current Engineering and Technology, vol. 2, no. 4, pp. 427–431, Dec. 2012.
32. A. Ray, and D. De, An energy efficient sensor movement approach using multi-parameter reverse glowworm swarm optimization algorithm in mobile wireless sensor network, Simulation Modelling Practice and Theory, vol. 62, pp. 117–136, 2016.
33. Q. Ni, H. Du, Q. Pan, C. Cao, and Y. Zhai, An improved dynamic deployment method for wireless sensor network based on multi-swarm particle swarm optimization, Natural Computing, vol. 16, no. 1, pp. 5–13, 2017.
34. R. Katsuma, Y. Murata, N. Shibata, K. Yasumoto, and M. Ito, Extending k-coverage lifetime of wireless sensor networks using mobile sensor nodes, in Proc. IEEE Int. Conf. Wireless Mobile Comput. Netw. Commun. (WIMOB), Marrakesh, Morocco, 2009, pp. 48–54.
35. S. Singh and A. Kumar, Novel optimal deployment of sensor nodes using bio inspired algorithm, in Proc. Int. Conf. Adv. Commun. Control Comput. Technol., Ramanathapuram, India, May 2014, pp. 847–851.

3

Metaheuristic Clustering in Wireless Sensor Networks

Sandip K. Chaurasiya, Suman Ghosh, and Rajib Banerjee

CONTENTS

DOI: 10.1201/9781003102397-3

3.1 Introduction

Due to its cost-effectiveness, simplicity, and ease of deployment, the wireless sensor network (WSN) has evolved as a very significant and promising technology. WSN enables a wide variety of applications serving the humanity (e.g., environmental monitoring, military surveillance, habitat monitoring, healthcare, public safety, and industry applications). Wireless sensor network comprises sensor nodes equipped with the ability to sense and measure the surroundings and to communicate the data to the base station either directly or using multi-hop for the access to the end user [1, 2]. However, the miniaturization of sensor nodes imposes a number of restrictions on the nodes, especially in terms of limited computational capacity, limited storage capability, and limited power. Among all these limitations, energy scarcity is the most severe one because it might not be feasible to replace the battery of the nodes after depletion. Hence, network operations in the wireless sensor network have to be energy efficient.

Since transmission at the physical layer and network layer consumes the most of the node's energy, energy-efficient data transmission in the WSN has attracted researchers a lot. As mentioned above, the nodes may communicate their data either directly to the base station or using the multi-hop approach. However, the distance between the nodes and base station matters the most in view of the energy consumption due to transmission. To address this issue, the clustering of nodes has been used. In clustering, nodes with some common attributes are grouped together [3]. More illustratively, in a cluster-based architecture, the network nodes are logically divided into a number of groups, each being called a cluster based on some criteria, viz. nodes' proximity, nodes' degree, nodes' residual energy, etc. Once the nodes are partitioned into clusters, one node in a cluster is designated as the cluster head (CH) which collects data from each of the cluster members, aggregates them, and finally forwards the aggregated data to the base station for further processing or end user access. The cluster heads may communicate their data to the base station directly or using multi-hop approach in which each CH forwards its data to another CH in the direction of base station. In this way, the localized coordination and the data aggregation along with the reduced communication overhead improve the overall network lifetime. It can be observed from the above discussion that multiple attributes can be brought into the consideration for the formation of clusters in the network, and hence the problem of cluster formation can be perceived as a multi-variable optimization problem. To address this class of problems, there are two popular approaches: heuristic and metaheuristic techniques. The heuristic technique is better known for its problem-dependent approach in which the complete set of problem particulars is exploited in search of an optimum solution. The metaheuristic technique is a problem-independent approach for solving the multi-objective nonlinear complex real-world problems.

The main motivation of this chapter is to explore the existing metaheuristic clustering solutions in order to figure out the appropriate future research direction to further address the aforesaid problem of energy-efficient clustering more effectively.

3.1.1 Major Contributions and Organization of the Work

The major contributions of this work are listed as follows:

- Illustration of the major components of the metaheuristic technique.
- Illustration of the major and popular variants of the metaheuristic technique.
- Thorough review of the popular metaheuristic-based contributions towards the energy-efficient clustering.

The rest of the chapter is organized in the subsequent five sections. In section 3.2, the notion of the metaheuristic technique is discussed in detail. Section 3.3 outlines the taxonomy of clustering in wireless sensor network. Section 3.4 discusses the existing contributions based on metaheuristic technique with an analytical view. Section 3.5 deals with the future research directions via metaheuristic approach and finally, section 3.6 concludes the entire work.

3.2 Metaheuristic Preliminaries and Related Concepts

As stated in the section 3.1, optimization schemes can be broadly categorized as heuristic and metaheuristic. The heuristic scheme exploits the complete set of problem-particulars in search of the most optimum solution; however, being a problem-dependent approach, heuristic algorithms might get trapped into local maxima/minima. Contrary to this, metaheuristic schemes are problem-independent and stochastically explore the search space to produce the most optimal solution without getting trapped into the local maxima/minima.

The general scheme of the metaheuristic technique is depicted in Figure 3.1 as follows: The scheme explores and exploits the available search space iteratively for a definite number of rounds. In addition to the two basic parameters – initial population and number of iterations for the function evaluation – every metaheuristic scheme may have some schematic control parameters, viz. mutation rate, crossover rate, scaling factor, etc., with algorithm-specific effects. The metaheuristic function is also enriched with a mandatory fitness function, which is used to evaluate the strength of the solution vectors appearing as the outcome of scheme. The higher the fitness value the better the solution.

The scheme starts with initializing a population of random solutions from the available search space. Every solution vector is then evaluated for its strength, indicating the suitability as the final solution through a well-defined fitness function. Afterwards, from the current population, the selected vectors are directed to undergo a variation phase that comprises mutation and crossover phases. The resultant vectors are evaluated for their strength, indicating the suitability of the tentative vectors as solutions. A survivor function is then called to select the vector for the next generation/iteration on the basis of the fitness values attained by the vectors before and after visiting the variation phase. The process

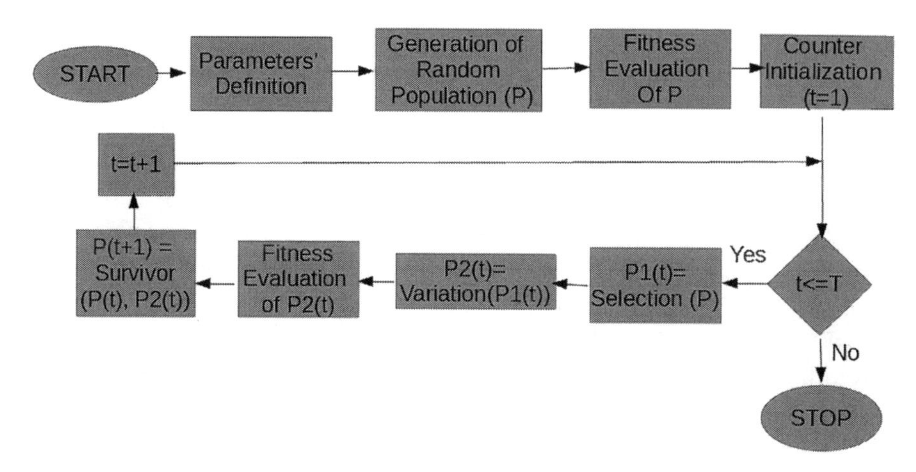

FIGURE 3.1
Meta-heuristic technique: A general scheme.

of updating the population vector continues for a definite number of times determined a priori, and finally the most recent solutions (i.e. the vectors after the last round) are taken as the final solutions.

From the above discussion, it can be explicitly concluded that the fitness function plays the most vital role in the performance of a metaheuristic scheme. If not designed carefully and intelligently, the fitness function may mislead the approach from its intended objective. Hence, be it any variant of metaheuristic scheme, viz. genetic algorithm, differential evolution, particle swarm intelligence, ant colony optimization, or teaching-learning-based optimization; one must design the fitness function with utmost care and as per the intended objective.

In this work, to illustrate the metaheuristic approach especially in the domain of wireless sensor network, three popular variants – differential evolution (DE) [4], particle swarm optimization (PSO) [5], and teaching-learning-based optimization (TLBO) [6] – are to be discussed in detail. The aforementioned schemes have been chosen because of their simplicity and faster rate of convergence towards the global optimum solution. Although DE and PSO have already been explored very well towards having energy-efficient network operations via metaheuristic clustering of sensor nodes in WSN, TLBO has not been explored towards the same. Authors have a strong belief that the TLBO will play a key role in the obtainment of metaheuristic clusters of the sensor nodes in the near future. Hereby, the readers are encouraged to experiment with the possibilities of obtaining load-balanced metaheuristic clusters in the wireless sensor networks using the unexplored teaching-learning-based optimization.

3.2.1 Differential Evolution

Proposed by Storn et al., differential evolution [4] has evolved as one of the most popular stochastic metaheuristic schemes over the continuous search space. The general scheme of differential evolution can be seen in the following Figure 3.2.

The scheme starts with the initialization of basic parameters – initial random population (say, of the size N) and number of iterations – and that of the schematic parameters-scaling factor and crossover rate. Once all the parameters are initialized, each of the solution

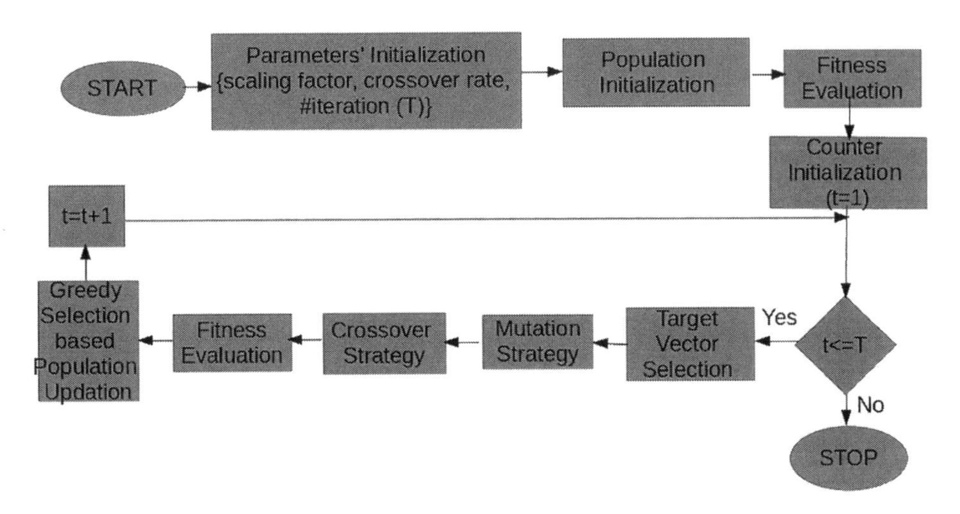

FIGURE 3.2
Differential evolution: A general scheme.

vectors (chromosome/genome) with D decision variables, known as target vector ($\Omega_i \mid i \in 1, 2, 3,. ., N$) from the initial population (Ω) is submitted to the mutation phase. There exists a wide variety of mutation strategies on the basis of choices made for differential vectors and their count, such as random vectors, best vector, and target-to-best vectors, etc. One can easily decide on and utilize the most suitable strategy for a specific problem. The target vector, once passed through the mutation phase, becomes the donor vector or mutant vector ($\Xi_i \mid i \in 1, 2, 3,. ., N$), where the complete set of mutant vector is represented by Ξ. After getting through the mutation phases, the target vector is submitted for the crossover beyond which the donor vector is termed the trial vector ($\Psi_i \mid i \in 1, 2, 3,. ., N$) and the set of such trial vectors is denoted by Ψ. The crossover strategy can be of the two types – binomial and exponential.

Binomial Crossover

$$\psi_j = \begin{cases} \xi_j & \text{if } r \leq C_r \text{ OR } j = \delta \\ \omega_j & \text{if } r > C_r \text{ AND } j \neq \delta \end{cases} \tag{3.1}$$

where, C_r is the crossover rate, δ is the randomly selected integer from the set {1, 2, 3,. . . , D}, r is the random number between 0 and 1, ψ_j refers to the jth variable of the trial vector, ξ_j refers to the jth variable of donor/mutant vector, and ω_j refers to the jth variable of the target vector.

Exponential Crossover: In the exponential crossover, at first, the nth variable from the donor vector is copied into the trial vector. Afterward, every subsequent variable from the donor vector is copied into the trial vector as long as the $r \leq C_r$. Once $r > C_r$, variables from the target vector are copied into the trial vector.

Depending on the chosen mutation and crossover strategies, a number of variants for the differential evolution have been proposed. In order to discriminate among those variants, Kendall's standard notation, *DE/x/y/z* is used. Here, DE indicates the differential

TABLE 3.1

Variants of the Differential Evolutions

DE Scheme	Mutation Strategy	Mutation Expression	Crossover Type
DE/rand/1/bin	Random	$\Xi = \Omega_{r1} + F.(\Omega_{r2} - \Omega_{r3})$	Binomial
DE/rand/2/exp	Random	$\Xi = \Omega_{r1} + F.(\Omega_{r2} - \Omega_{r3}) + F.(\Omega_{r4} - \Omega_{r5})$	Exponential
DE/best/1/bin	Best	$\Xi = \Omega_{best} + F.(\Omega_{r1} - \Omega_{r2})$	Binomial
DE/best/2/bin	Best	$\Xi = \Omega_{best} + F.(\Omega_{r1} - \Omega_{r2}) + F.(\Omega_{r3} - \Omega_{r4})$	Binomial
DE/target-to-best/1/exp	Target-to-Best	$\Xi = \Omega_i + F.(\Omega_{best} - \Omega_i) + F.(\Omega_{r1} - \Omega_{r2})$	Exponential
DE/target-to-best/2/exp	Target-to-Best	$\Xi = \Omega_i + F.(\Omega_{best} - \Omega_i) + F.(\Omega_{r1} - \Omega_{r2}) + F.(\Omega_{r3} - \Omega_{r4})$	Exponential

evolution, x refers to the selected mutation strategy, y denotes the number of differential vectors being used in the mutation scheme, and z refers to the adopted crossover strategy. A few major variants being popularly used are listed in Table 3.1.

In Table 3.1, $F \in (0, 2)$ is the scaling factor, Ω_i is the ith target vector, best refers to the target vector with the best fitness value, and Ω_{r_j} is the jth target vector chosen randomly where $j \in [1, N]$, N being the number of target vectors in the population i.e. the population size. Once all the trial vectors are generated for the target vectors of current generation, say G, descendent offspring are decided based on the fitness values of comparable pairs of target and trial vectors i.e. $< \Omega_{i,G}, \Psi_{i,G} >$ for $i \in [1, N]$ as follows:

$$\Omega_{i,G+1} = \begin{cases} \Psi_i & \text{if } fitness(\Psi_{i,G}) \geq fitness(\Omega_{i,G}) \\ \Omega_i & \text{otherwise} \end{cases} \tag{3.2}$$

The entire scheme of the differential evolution can also be summarized in the following algorithm.

Algorithm – DE

Input:

- Ω_N: Initial Population of size N
- T: No. of iteration
- ff^n (): Fitness function
- F: Mutation/Scaling factor
- C_r: Crossover rate

1. BEGIN
2. Evaluate fitness of Ω_N i.e. $ff^n(\Omega_N)$
3. for t $\leftarrow$ 1 : T
4.   for i $\leftarrow$ 1 : Ω_N
5.   Generate the donor vector (Ξ_i) using mutation
6.   Perform crossover to generate offspring (Ψ_i)
7.   end for
8.   for i $\leftarrow$ 1 : Ω_N

9. Evaluate the fitness of Ψ_i i.e. $ff^n(\Psi_i)$
10. Greedy selection between Ω_i & Ψ_i based on their respective fitness value
11. Update the population
12. end for
13. end for

3.2.2 Particle Swarm Optimization

Developed by Kennedy et al., particle swarm optimization [5] is a guided random search algorithm that is inspired by the collective behavior of social animal societies. However, it usually models the social behavior of fish schooling or birds flocking. In this technique, every particle is characterized by a position and a velocity associated with it. The particles may change their respective position by tuning their velocity. The particles are also equipped with the capability to remember the best position achieved by it and can communicate the best position explored by them to their peers. Like in the general scheme of metaheuristic techniques, the initial position and velocity population vectors (say, Ω and Θ respectively) are generated randomly but within the available search space. After such random initialization of preliminary populations, the fitness of each position vector is computed. It is to be noted that if the initial population referring to the initial positions of the particle is of the size N, each with D decision variables, then Ω would be an $[N \times D]$ matrix whereas the fitness-matrix (say, Γ) referring to the fitness values of position vectors would be an $[N \times 1]$ matrix. Along with the maintenance of the Ω, Θ, & Γ, two temporary matrices-Ω_{best} and Γ_{best} are also maintained in the PSO for the determination of the final solution vector. Here, Ω_{best}, initially being same as that of Ω, keeps track of the best position explored by the vectors themselves; hence known as **personal best** and Γ_{best} maintains the corresponding values attained by the vectors in best. From the best, a vector is chosen with the best possible fitness value, which is termed the **global best** denoted by Λ_{best}.

The scheme proceeds in a way such that in every round, each velocity vector, say Θ_i of Θ, is processed as per the following equation:

$$\Theta_i = \omega.\Theta_i + c_1.r_1.(\Omega_{best,i} - \Omega_i) + c_2.r_2.(\Lambda_{best} - \Omega_i) \quad (3.3)$$

where, $\omega \in (0, 1)$ is the particle-inertia, c_1 & c_2 are the acceleration coefficients $\in (0, 2)$, and r_1, r_2 are random row vectors of the length D. Once Θ_i is obtained as per the (3), the corrspoding position vector say Ω_i from Ω is also updated as follows:

$$\Omega_i = \Omega_i + \Theta_i \quad (3.4)$$

After the obtainment of Ω_i, the population vector is updated irrespective of its current fitness value (i.e. no greedy approach is followed for updating the Ω); however, Ω_{best}, Γ_{best}, and Λ_{best} are updated following the greedy approach as follows:

$$\Omega_{best,i} = \Omega_i \;\; if \;\; ff^n(\Omega_i) < ff^n(\Omega_{best,i}) \quad (3.5)$$

And,

$$\Lambda_{best} = \Omega_{best,i} \;\; \text{if} \;\; ff^n(\Omega_{best,i}) < ff^n(\Lambda_{best}) \quad (3.6)$$

After executing the aforesaid steps for a definite number of rounds, one may have the final solution.

Moreover, the entire scheme of the particle swarm optimization can also be summarized as the following algorithm.

Algorithm – PSO

Input:

- Ω: Initial positions of the particles
- Θ: Initial velocities of the particles
- T: No. of iteration
- ff^n (): Fitness function
- w: Particle inertia
- $c_1 \& c_2$: Acceleration coefficients
- $\Omega_{best,i}$: Personal best of the ith particle
- Λ_{best}: Global best of the ith particle
- Ω_i: Position of the ith particle from the population Ω

1. BEGIN
2. Evaluate fitness of Ω i.e. $ff^n(\Omega)$
3. $\Omega_{best} = \Omega$
4. $\Lambda_{best} = best - fitness - value(\Omega_{best})$
5. for t $\leftarrow$ 1 : T
6.   for i $\leftarrow$ 1 : $||\Omega||$
7.   Determine the velocity Θ_i of the ith particle as in (3) i.e.

$$\Theta_i = \omega.\Theta_i + c_1.r_1.(\Omega_{best,i} - \Omega_i) + c_2.r_2.(\Lambda_{best} - \Omega_i)$$

// Here, r_1 & $r_2 \rightarrow$ random numbers' vectors of size $[1 \times D]$

8.   Determine the position (Ω_i) for the ith particle as in (4) i.e.

$$\Omega_i = \Omega_i + \Theta$$

9.   Evaluate the fitness function for the ith particle of Ω
10.   Update the Ω by including Ω_i
11.   Update the $\Omega_{best,i}$ if applicable
12.   Update the Λ_{best} if applicable
13.   end for
14.   end for
15. END

3.2.3 Teaching-Learning-Based Optimization

Developed by Rao et. al., the teaching-learning-based optimization [6] is a metaheuristic technique inspired by how the knowledge is transferred in a regular class. In comparison to the two above-mentioned schemes, TLBO requires only one schematic parameter – teaching factor (T_f) – in addition to the basic parameters – initial population (Ω) and number of round (T). The scheme comprises two phases – teacher phase and learner phase. Each round consists of teacher phase and a learner phase as explained in Figure 3.3.

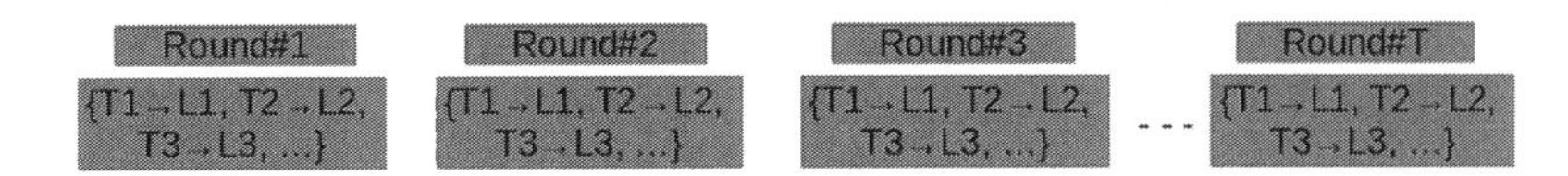

FIGURE 3.3
Working of TLBO over the rounds.

Each solution vector from Ω undergoes the teacher phase followed by the learner phase in each round. The working of the TLBO is explained below:

3.2.3.1 Teacher Phase

In the teacher phase, the vector with the best fitness value, say Ω_{best} from Ω, is chosen and the mean of the vectors $\in \Omega$, say Ω_{mean}, is computed. Thereafter, the new solution is derived on the basis of Ω_{best} and Ω_{mean} as follows:

$$\Omega_{new} = \Omega_i + r.\left(\Omega_{best} - T_f.\Omega_{mean}\right) \tag{3.7}$$

where, T_f is known as the teaching factor, which can assume either 1 or 2 as its value; Ω_i, Ω_{best}, Ω_{mean}, Ω_{new} are the current, best, average, and new population vectors under processing. After obtaining Ω_{new}, its fitness value is computed and compared with that of Ω_i in order to decide whether to update the Ω_i. That is, if the fitness value of the Ω_{new} is better than the fitness value of Ω_i, then Ω is updated with the newly computed Ω_{new}; otherwise the previous value Ω_i persists.

3.2.3.2 Learner Phase

In this phase, the solution obtained from the respective teacher phase is improvised further. Let the solution from the previous teacher phase be denoted by Ω_i; then a random solution is selected from the Ω other than Ω_i which is called partner solution (Ω_p). Now, this randomly selected partner solution is processed as per following equation to derive the new solution, Ω^i_{new}.

$$\Omega^i_{new} = \begin{cases} \Omega_i + r.\left(\Omega_i - \Omega_p\right) & \text{if } ff^n\left(\Omega_i\right) < ff^n\left(\Omega_p\right) \\ \Omega_i - r.\left(\Omega_i - \Omega_p\right) & \text{if } ff^n\left(\Omega_i\right) \geq ff^n\left(\Omega_p\right) \end{cases} \tag{3.8}$$

As used previously, here, $ff^n\left(\Omega_i\right)$ refers to fitness value of Ω_i and similary the same is true for $ff^n\left(\Omega_p\right)$

Thus obtained Ω^i_{new} is now compared with the existing Ω_i and the one with greater fitness value is used update the Ω

The steps elaborated above are repeated for the round specified a priori, and after the last run, one achieves the final solution. The idea of the teaching-learning-based optimization can also intuited with the help of algorithm detailed below:

Algorithm – TLBO
Input:

- Ω: Initial positions of the particles
- T: No. of iterations

- $ff^n()$: Fitness function
- Ω_{best}: Vector with the best fitness value in Ω
- Ω_{mean}: Mean of the vectors from the population Ω

1. BEGIN
2. Evaluate fitness of Ω i.e. $ff^n(\Omega)$
3. for t $\leftarrow$ 1 : T
4. for i $\leftarrow$ 1 : $||\Omega||$
5. Determine the Ω_{best} & Compute Ω_{mean} for population Ω
6. Compute the new solution vector Ω_{new} as

$$\Omega_{new} = \Omega_i + r.\left(\Omega_{best} - T_f.\Omega_{mean}\right)$$

// Here, $r \rightarrow$ random numbers' vector of size $[1 \times D]$

7. if $ff^n(\Omega_{new}) < ff^n(\Omega_i)$
   - Update Ω with Ω_{new}
8. else
   - Continue with Ω_i
9. Choose a solution, Ω_p from Ω such that $p \neq i$

// Ω_p is referred to as Partner Solution

10. if $ff^n(\Omega_i) < ff^n(\Omega_p)$
    - $\Omega_{new} = \Omega_i + r.\left(\Omega_i - \Omega_p\right)$
11. else
    - $\Omega_{new} = \Omega_i - r.\left(\Omega_i - \Omega_p\right)$
12. if $ff^n(\Omega_{new}) < ff^n(\Omega_i)$
    - Update Ω with Ω_{new}
    - Update Ω_{best} & Ω_{mean}
13. else
    - Continue with Ω_i
14. end for
15. end for
16. END

3.3 Clustering in WSN

This section discusses the terminology and the various attributes used in the analysis of clustering strategy as follows:

3.3.1 Clustering Environment

The clustering environment refers to how the nodes are grouped together to shape the clusters. Based on the approach followed, wireless sensor networks can be centralized, distributed, or hybrid. In the centralized clustering, a comparatively wider set of node-specific information is required to decide upon the clusters. Either the base station (BS) or the concerned sensor node in the network controls the entire clustering process by exploiting the global information by the network nodes. The centralized clustering curbs the network performance and lowers the network efficiency. Contrary to this, in the distributed approach, the network nodes coordinate with one another in order to decide upon the responsibility (head node/member node) in the network. The nodes collaboratively take various network decisions, viz. joining the cluster, building the date-routes, etc. The distributed clustering contributes to energy-efficient, scalable, and improved network operations. The merits of both the aforesaid approaches can be combined in the hybrid approach. The centralized approach may result in effective network partitioning, and then the distributed approach can be applied for cluster heads coordination [7].

3.3.2 Clustering Mode

Clustering mode refers to how frequently the clusters change their shapes/configurations. There are three approaches – static, dynamic, and hybrid. In the static mode, once the nodes are configured to shape clusters, the clusters formed will not change with time [8, 9]. More illustratively, once the nodes are partitioned into clusters, the clusters do not change and remain static throughout the network lifetime. On the other hand, in the dynamic clustering, the clusters change with the network round (i.e., the sensor nodes are configured together afresh to form the clusters in every network round) [10, 11]. However, the repetitive clusters' formation may consume a lot of network energy. The hybrid approach has the benefits of both the aforesaid approaches. The static mode can be applied for the effective network-partitioning, saving energy-consumption from the repetitive clusters' formation, and the dynamic approach can be applied for even load distribution among the network nodes via cluster heads rotations, etc.

3.3.3 Clustering Characteristics

There are various attributes associated with the clusters, viz. cluster size, cluster count, and cluster communication. Cluster size indicates the number of nodes participating in the clusters. The size of the clusters may equal or unequal. Depending on the nature of application, one may choose the suitable strategy to proceed ahead with the clustering. Similarly, the cluster count refers to the number of clusters required for the network operation. The cluster count can be either fixed [8, 9] or variable [10, 11].

Cluster communication describes the message exchange among the network nodes. It can also be categorized in two broader categories – intra-cluster and inter-cluster communication. The intra-cluster communication refers to the message exchange between the member nodes and their respective cluster heads; inter-cluster communication indicates the message exchange among the cluster heads. In inter-cluster communication, each of the communicating heads deals on behalf of its respective cluster.

3.3.4 Clustering Process

The main objective of the clustering process is the obtainment of energy-efficient network operations to facilitate improved lifetime in addition to even load distribution, fault tolerance, and scalability, etc. Mainly, the clustering comprises two phases –

formation of the clusters and selection of cluster heads. In the cluster formation phase, nodes join the cluster heads basically on the basis of the received signal strength. In the cluster-head selection phase, a specially designated node called the cluster head is selected. The cluster head is the node that is responsible for the energy-intensive network operations, viz. data reception from the multiple sensor nodes, data aggregation, and transmission to the base station. However, the selected cluster head might be variable (i.e., periodic rotation of the role of the cluster head among the cluster member) [8, 9] or might remain the fixed throughout the network lifetime, depending on the nature of application [12].

3.3.5 Clustering Objectives

Wireless sensor networks are deployed to serve a specific application and hence are called application-specific networks. Every application comes with its own specific requirements, classifying the objectives/purposes of the deployed network. It has been proven that the clustering of the nodes is a significant tool for the attainment of these objectives. The clustering objectives can be categorized into two main classes – explicit and implicit objectives as indicated in the Figure 3.4.

As implied by the names, explicit objectives are the ones that can be observed as the direct outcomes of the clustering process; the implicit objectives are the indirect or secondary consequences of the clustering process.

More illustratively,

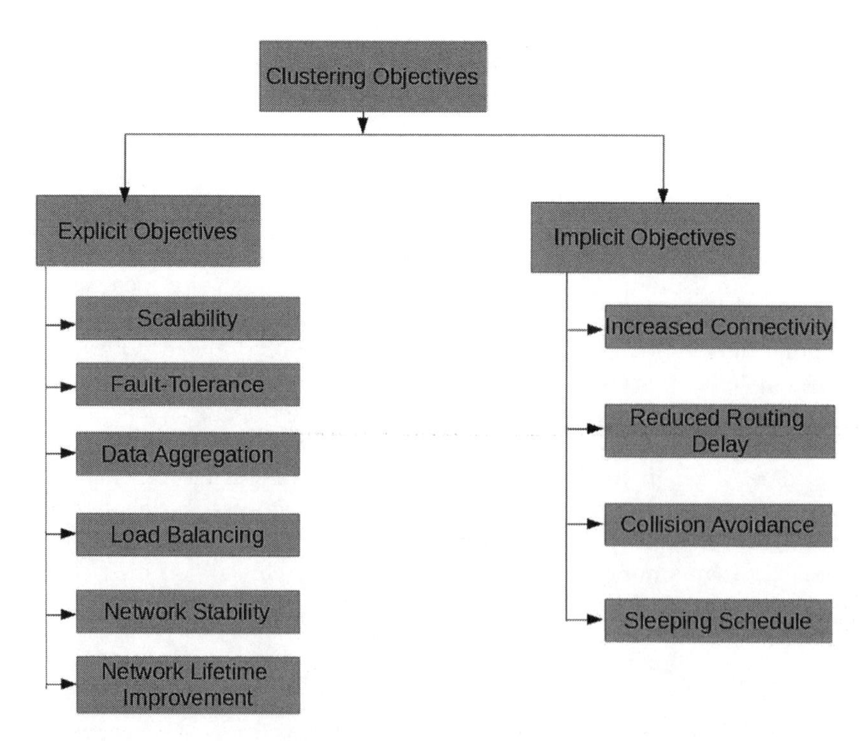

FIGURE 3.4
Clustering objectives in wireless sensor networks.

3.3.5.1 Explicit Clustering Objectives

- **Scalability**
  Scalability indicates the extent to which the proposed solution is independent of the network size. In other words, it refers to the ability of the solution to perform consistently well irrespective of the number of nodes in the network (i.e., with the increase or decrease in the deployed nodes, the solution doesn't underperform). Clustering in WSNs supports this idea, as the nodes can be added or deleted in a cluster without much computational overhead.
- **Data Aggregation**
  In the regular deployment of sensor nodes, nodes might get deployed closely to one another. Due to the nodes' proximity, data generated by the nearby nodes are redundant. Clustering answers this data redundancy in a way that the cluster head collects the data from its member nodes, normalizes the data while removing the redundancy present, and generates an aggregated version of the data packets received.
- **Fault Tolerance**
  Fault tolerance refers to the ability of the network to sustain operations even in the presence of faults (i.e., even though a few nodes are dead in the network, the network is performing satisfactorily). This must be targeted in the networks, such as a sensor network, as the majority of deployments are humanly inaccessible. Clustering of the nodes enables fault tolerance in such a way that failure of one or a few nodes does not compromise the purpose of network deployment. Since the nodes in the cluster are close to one another, they are supposed to generate spatially highly correlated data. Due to the virtue of this high spatial correlated data, even though a few nodes stop working for some reasons, viz. battery depletion, etc., the data generated by the other live nodes might result in an aggregated data packet as if any node-failure had not happened.
- **Load Balancing**
  Inside the cluster, the normal member nodes perform only low-energy-consuming tasks, such as sensing the surroundings, processing the collected data only to a limited extent, and handing over the respective data to the cluster head. However, the cluster head is responsible for data aggregation and long-range data transfer to the base station, which are obviously high-energy-consuming tasks. In order to distribute the load evenly among the nodes, clustering provides rotation of the role of cluster head. Such rotation of the role of cluster head within the clusters facilitates in achieving an energy-efficient prolonged network.
- **Topological Stability**
  Topological stability refers to the management of the nodes in the network. Within a cluster, the cluster head is usually aware of the node-specific information of its member nodes, viz. residual energy, location, and node-degree, etc. With the help of this information, the cluster head can easily report any change in the topology if detected due to nodes' death or nodes' movement.
- **Network Lifetime Improvement**
  Network lifetime can be defined as the time when either the first node dies or the last node dies. The foremost objective of the clustering is the achievement of energy-efficient network operations, and the energy-efficiency is achieved by the aforementioned even load distribution, cluster-head rotations within the clusters,

aggregation, etc. Such energy-efficient network operations result in prolonged network lifetime as the energy saved due to the aforesaid operations can be used in further network actions such as data transmission, etc. The network lifetime can be further improved by implementing some additional policies, viz. selection of centrally placed nodes as cluster heads, minimizing the intra-cluster communication cost, associating subordinate nodes to the cluster heads, and applying effective sleeping strategies in the clusters.

3.3.5.2 Implicit Objectives

- **Enhanced Connectivity**
  Compared to the flat architecture, improved connectivity is assured in the clustered network. Clustering enables the network to be attributed with a *k-connected network* where *k* is the number of nodes in a cluster. The *k-connected network* ensures the livelihood of the network until all the k nodes are dead in comparison to at networks where it is expected to be in communication with the base station for every node in the network.
- **Reduced Routing Delay**
  Compared to the at network architecture in which multi-hop communication is traditionally implemented, the clustering process proceeds with single-hop communication preferably. In time-sensitive applications where delay matters a lot, the traditional multi-hop approach in at network architecture may be the least required. In the worst case, the length of path following the multi-hop approach can be of the order of n, where n is the number of nodes; in the case of clustered network architecture, the maximum delay can be of the order k, where k is the number of clusters. Thus, the clustering of nodes also ensures minimized delay.
- **Collision Avoidance**
  In densely deployed networks, viz. wireless sensor networks, collision of data packets by the nodes is a major issue. It may result in wastage of network energy, as the nodes have to retransmit the lost packets to meet the requirement of the application. This huge wastage of nodes' energy in the retransmission of packets due to collision can be avoided by applying some specialized MAC layer protocols – time division multiple access (TDMA) and carrier sense multiple access (CSMA) – within the clusters.
- **Sleeping Schedule**
  In networks like wireless sensor networks, where a huge number of nodes are densely deployed to one another, it is always advisable not to keep radio nodes active all the time. Instead, the radio nodes are activated and deactivated as per a schedule, which is known as the sleeping schedule. Implementing the sleeping schedule in the clustered network is quite simple as the cluster heads may easily define the schedule for their respective member nodes.

3.3.6 Characteristics of Cluster Heads

Being the most important part of a cluster, cluster heads get significant consideration in the development of network solutions; and hence, the characteristics of cluster heads, viz. mobility, ability, selection, and role, are crucial.

3.3.6.1 Selection of the Cluster Head

There are three major methodologies for the selection of cluster heads – preset selection, random selection, and attributed selection. As implied by the name, preset selection of the cluster heads refers to the cluster heads that are deterministically deployed for this very purpose as in [12] where gateway nodes are deployed to serve the purpose of cluster heads only. Random selection as in [10] figures out the heads for the respective clusters randomly. Contrary to random selection, in the attributed selection of cluster heads, nodes' status in terms of their location, residual energy, proximity, etc. are used in determining a suitable cluster head for a cluster as in [7–9].

3.3.6.2 Responsibilities of the Cluster Head

The main responsibilities of the cluster heads are to perform aggregation of the data received from their respective member nodes and to transmit the aggregated data to the base station using either the single-hop approach or via multi-hop. In single-hop data transmission, cluster heads send their data directly to the base station; in the multi-hop communication, other cluster heads present in the network are used while minimizing the path-cost to base station.

3.3.6.3 Mobility of the Cluster Head

Cluster heads can be stationary or mobile as per the need of application. However, topology management in case the cluster heads are mobile becomes a little difficult in comparison to having stationary nodes.

3.4 Metaheuristic Clustering Approaches

Over the past few years, metaheuristic techniques have been deployed along with clustering in order to devise the energy-efficient and optimized network operations. Several such techniques have been used successfully for the aforesaid. As stated in the prologue of section 3.2, three major variants of the metaheuristic approach – differential evolution [4], particle swarm optimization [5], and teaching-learning-based optimization [6] are discussed in detail here.

3.4.1 Differential Evolution-Based Clustering Approaches

As explained in the section 2.1, the differential evolution (DE) scheme involves two main steps – mutation and crossover. Table 3.1 lists the major variants of the mutation strategies and the crossover techniques that can be combined according to the researcher's discretion in order to have optimized network solutions. In addition to the aforementioned steps, having an appropriate fitness function to measure the suitability of the tentative solutions is a must for the efficacy of the scheme. Hereby, derivation of an objective-specific fitness function has also been addressed equally in most of the solutions proposed by the research community.

A large number of contributions have already been proposed based on this outstanding evolutionary technique (i.e., differential evolution in search of suitable clusters of nodes in WSN). This subsection discusses some of the prime contributions in this regard as follows:

X. Li et al. proposed a differential-evolution-based routing scheme for environmental monitoring wireless sensor networks, DE-LEACH [13]. DE-LEACH applies the simple and quickly converging differential evolution in order to produce optimized clusters by considering two parameters – residual energy information of nodes and spatial distribution of nodes. The scheme comprises four phases, namely partitioning initial clusters, collecting status information of the nodes within the clusters through the auxiliary cluster heads, determining optimized cluster heads with differential evolution, and forming optimized clusters. The phases are to be executed in every round of the network operation.

The scheme starts with the clusters formed as in the traditional LEACH [10], which are then optimized by applying the *DE/rand/*1 mutation strategy along with the binomial crossover scheme based on the information collected about the member nodes by the auxiliary cluster heads (cluster heads selected as per the scheme defined in [10]) accordingly. The fitness function used to evaluate the strength of the ith vector in the population is defined as

$$fitness(i) = \frac{1}{N}.\sum_{j=1}^{N}\left(\alpha.RE_j^i + \beta.AvgRE_j^i\right) \tag{3.9}$$

where α & β are defined such that $\alpha, \beta \in [0, 1]$ and $\alpha + \beta = 1$; RE_j^i refers to the residual energy of the jth component node of the ith vector; and $AvgRE_j^i$ is the average energy of all the nodes in the ith vector except the jth component node.

The scheme outperforms the traditional LEACH and LEACH-C [11]. However, heavy computational responsibilities are imposed on the nodes.

P. Kuila et al. proposed a differential-evolution-based clustering algorithm (DECA) [12] that deploys some specialized nodes for the role of cluster heads. These special nodes are called relay nodes or gateway nodes and are equipped with additional initial energy to execute energy-intensive tasks such as long-distance transmission to the sink and so on. DECA utilizes *DE/best/*1 as its mutation strategy along with the binomial crossover. In addition to this, a new local improvement phase is also proposed in DECA to facilitate the even load distribution among the nodes to ensure improved longevity of the nodes in the network.

The fitness function used in the scheme takes into account the standard deviation of lifetime of gateways ($\sigma_{Lifetime}$) and average cluster distance ($\sigma_{Distance}$) in its derivation as follows:

$$fitness \propto \frac{1}{\sigma_{Lifetime}}.\frac{1}{\sigma_{Distance}} \tag{3.10}$$

Here,

$$\sigma_{lifetime} = \sqrt{\frac{1}{k}\sum_{j=1}^{k}\left(\mu_{Lifetime} - Lifetime(j)\right)^2} \tag{3.11}$$

where k is the number of gateways deployed; *Lifetime*(j) is the lifetime of the jth gateway; & $\mu_{Lifetime} = \frac{1}{k}\cdot\sum_{i=1}^{k} Lifetime(i)$.

And,

$$\sigma_{Distance} = \sqrt{\frac{1}{k}\sum_{j=1}^{k}\left(\mu_{Distance} - ACD(j)\right)^{2}} \tag{3.12}$$

where k is again the number of gateways deployed; $ACD(j)$ is average distance of the jth gateway from the base station; and $\mu_{Distance} \frac{1}{k} \cdot \sum_{i=1}^{k} .ACD(i)$.

Hereby, from equations (3.10)–(3.12),

$$fitness = \frac{k}{\sqrt{\frac{1}{k}\sum_{j=1}^{k}\left(\mu_{Lifetime} - Lifetime(j)\right)^{2}} \cdot \sqrt{\frac{1}{k}\sum_{j=1}^{k}\left(\mu_{Distance} - ACD(j)\right)^{2}}} \tag{3.13}$$

wherein, "K" being the proportionality constant can be substituted with 1 without any loss of generality. And hence,

$$fitness = \frac{1}{\sqrt{\frac{1}{k}\sum_{j=1}^{k}\left(\mu_{Lifetime} - Lifetime(j)\right)^{2}} \cdot \sqrt{\frac{1}{k}\sum_{j=1}^{k}\left(\mu_{Distance} - ACD(j)\right)^{2}}} \tag{3.14}$$

The scheme outperforms the [14–16], traditional DE (i.e., DECA without local improvement phase), and genetic algorithm-based schemes in terms of network lifetime; however, the scheme gives little attention to the cluster balancing only via its local improvement phase.

S. Potthuri et al. proposed a hybrid differential evolution and simulated annealing (DESA) [17] scheme with an objective to improve network lifetime in wireless sensor network. The scheme applies a hybrid of the DE and simulated annealing (SA) for local and global optimal solutions respectively. The scheme comprises four phases – population vector initialization, mutation, crossover, and selection as in the traditional differential-evolution-based strategy. However, for the initialization of population vectors, a more effective, "opposite point method" [18] is used instead of the random selection of population vectors. The mutation strategy is chosen dynamically based on the generation of a random number. A random number $\in (0, 1)$ is chosen, and if it comes below a predetermined threshold value (0.5 in the DESA), $DE/rand/1$ is chosen as the mutation scheme; otherwise, it's $DE/target - to - best/1$. And for crossover, a blending rate (B_r) based on Gaussian distribution is used as follows.

$$\psi_{j,i,G+1} = B_r * \omega_{j,i,G} + (1 - B_r) * \xi_{j,i,G} \tag{3.15}$$

where $\omega_{j,i,G}$ & $\xi_{j,i,G}$ are the jth component of the ith target and donor vector in the Gth generation; $\psi_{j,i,G+1}$ is the jth component of the ith trial vector in the $(G + 1)^{th}$ generation; and, B_r is defined below:

$$B_r = N\left(0.5, (1/2\pi)\right)$$

with $N(\mu, \sigma)$ being the normal distribution with mean, μ and standard deviation, σ.

DESA employs a fitness function which considers the ratio of nodes' energy to that of the respective clusters as follows:

$$fitness = \varepsilon * f_1^n + 1(1 - \varepsilon) * f_2^n \tag{3.16}$$

where ε is a user defined constant determining the contributions of f_1^n & f_2^n defined below:

$$f_1^n(i) = \frac{RE_i}{\sum_{k=1,k\neq i}^{m} RE_k}$$

$$f_2^n(i) = \frac{(m-1)}{\sum_{k=1,k\neq i}^{m} distance(i,k)}$$

here, RE_i is the residual energy of the ith node in the concerned cluster and *distance*(*i*, *k*) is the Euclidian distance between the ith and k^{th} node.

The scheme outperforms the traditional DE scheme in terms of throughput, energy consumption, and network lifetime etc.; however, it converges slowly.

M. Gahramani and Abolfazl Laakdashti proposed a scheme of efficient energy consumption in wireless sensor networks using an improved-differential-evolution algorithm [19]. The scheme by [12] is improved by updating the mutation strategy to accommodate the target vector along with the vector with best fitness value and two random population vectors as follows:

$$\Xi_{i,G} = \Omega_{best,G} + R_1.(\Omega_{i.G} - \Omega_{r_1,G}) + R_2.(\Omega_{r_2,G} - \Omega_{i,G}) \tag{3.17}$$

where R_1 & R_2 are the two random numbers such that $R_2 = 1 - R_1$; Afterwards, [19] utilizes binomial crossover in order to devise the trial vectors (Ψ). In addition to this, fitness function used by [19] improvises the fitness function used by [12] in such a way that it accommodates the total energy of the gateways and nodes in addition to the existing network lifetime standard deviation component as follows:

$$fitness = \alpha * \sum_{j=1}^{k} RE_{G_j} + \beta * \sum_{i=1}^{N} RE_{node_i} + \sqrt{\frac{1}{k} * \sum_{j=1}^{k} (\mu_{Lifetime} - Lifetime(j))^2} \tag{3.18}$$

where k is the number of gateways deployed and N is the total number of nodes; RE_{G_j} & RE_{node_i} are the residual energies of the jth gateway and i^{th} node respectively; $\mu_{Lifetime}$ (standard deviation of the lifetime of gateways) and $Lifetime_j$ (lifetime of the jth gateway) are defined as follows:

$$\mu_{Lifetime} = \frac{1}{k} \sum_{j=1}^{k} Lifetime_j$$

$$Lifetime_i = \left[\frac{RE_{G_i}}{GW_{Consumption}(G_i)} \right]$$

here, $GW_{Consumption}(G_i)$ is the total energy consumption by the ith gateway defined as

$$GW_{Consumption}(G_i) = N_i * E_{RX} + N_i * E_{DA} + E_{TX}(G_i, BS)$$

E_{RX} is the energy consumed in receiving a data packet; E_{DA} is the energy requried for aggregating the contents of a data packet; & $E_{TX}(G_i, BS)$ is the energy requried by the gateway in transmitting the data packet to the base station.

However, in the scheme [19], the problem of load balancing has not been addressed adequately.

M. Kaddi et al. proposed a metaheuristic scheme, a differential-evolution-based clustering routing protocol (DEBCRP) [20] for the wireless sensor network. The proposed scheme in [20] is a centralized scheme in which the base station coordinates the actions within the network. The scheme utilizes the *DE/rand/*1 mutation strategy along with the binomial crossover for the determination of appropriate clusters within the network. DEBCRP derives a fitness function which considers the nodes' residual energies with respect to the probable cluster heads and the distance between the nodes and the cluster heads for the formulation of clusters as follows:

$$fitness = \alpha * f_1^n + (1-\alpha) * f_2^n \tag{3.19}$$

where α is a user defined constant defining the impact of sub-objectives- f_1^n, & f_2^n; f_1^n, and f_2^n are defined as,

$$f_1^n = \sum_{i=1}^{k} \frac{\sum_{j=1, j \neq i}^{m} RE_j}{RE_{CH_i}}$$

$$f_2^n \sum_{i=1}^{k} \sum_{j=1, j \neq i}^{m} distance\ (j, i)$$

here, k refers to the number of cluster heads known a priori; m denotes the concerned cluster size; RE_i denotes the residual energy of the ith node; RE_{CH_i} denotes the residual energy of the ith cluster head; and *distance*(j, i) refers to the distance between the jth member of ith cluster head.

Once the clusters are formed, cluster heads collect data from their respective cluster members and aggregate the received data. However, in order to transmit the aggregated data to the sink, instead of following the single hop communication style, a PEGASIS [21] like chain of cluster heads is formed. DEBCRP outperforms the S-DE [22] in terms of network lifetime.

In [20], formulation of load-balanced clusters was not handled adequately. Also, a PEGASIS-like chain of the cluster heads suffers with similar problems as in [21], viz. delayed communication and inaccuracy of data due to aggregation of data by multiple cluster heads.

3.4.2 Particle Swarm Optimization-Based Clustering Approaches

As stated earlier in section 2.2, particle swarm optimization is a guided random search scheme that imitates and applies the collective behavior of social animal societies, viz. fish schooling and birds flocking. In the PSO, each particle is characterized by a set of two attributes – velocity and the position associated with it. Equations (3) and (4) define the way to update the velocity and position of the particles' vector respectively. Moreover, designing a meaningful and objective specific fitness function is a key for the success of the scheme.

A huge contribution has already been made towards obtaining the energy-efficiency in wireless sensor networks using the aforesaid particle swarm optimization techniques. This subsection discusses some of the contributions pertaining to energy-efficient clustering in the wireless sensor networks.

Latiff et al. proposed an energy-aware clustering for wireless sensor networks using particle swarm optimization [22]. The authors have attempted developing a centralized and energy-aware clustering protocol for the improvement of network lifetime of WSN. [22] focuses on producing load-balanced clusters with an objective to minimize the cost of intra-cluster communication among the network nodes while utilizing the PSO as the underlying scheme. The scheme comprises two phases – setup phase and steady state phase. As implied by the name, the setup phase being executed by the base station calls for the PSO to formulate the k-optimized clusters on the basis of node-specific information, residual energy (RE), and the location. Once the clusters are formed, the steady state phase calls for the action and data from the field to be transported to the base station for further processing.

[22] derives a fitness function that considers the distance of the furthest cluster member nodes from the respective cluster heads and the residual energy of the cluster heads as follows:

$$fitness = \beta * f_1^n + (1-\beta) * f_2^n \tag{3.20}$$

where β is a user-defined constant used to weigh the impact of the sub-objectives, f_1^n, f_2^n defined below:

$$f_1^n = \max_{i=1,2,3,\dots,k}\left[\sum_{\forall Node_j \in Cluster_{p,i}} \frac{distance\left(Node_j, CH_{p,i}\right)}{\left|Cluster_{p,i}\right|}\right]$$

$$f_2^n = \frac{\sum_{i=1}^{N} RE_i}{\sum_{j=1}^{k} RE_{CH_{p,j}}}$$

here, $Cluster_{p,i}$ refers to the $Cluster_i$ of the particle p; $\left|Cluster_{p,i}\right|$ denotes the number of nodes in the $Cluster_i$ of the particle p; and rests are as per the previously discussed style in the text.

The aforesaid scheme outperforms the [10] & [11] in terms of network lifetime.

Bennani et al. proposed particle swarm optimization-based clustering in wireless sensor networks: The effectiveness of distance altering [23]. In [23], a centralized scheme is discussed in which the base station plays a key role in deciding the clusters in the network. The network operations are divided into rounds in which every round comprises two phases – setup phase and steady state phase. In the setup phase, the network is partitioned into a number of clusters by the base station by employing the particle swarm optimization of the information collected from the deployed sensor nodes. Once the clusters are decided along with the respective cluster heads, periodic data transfer happens in the steady state phase.

In order to formulate the optimal clusters while distributing the energy load evenly among the network nodes, the fitness function was designed by considering the distance of the cluster heads from the base station along with the intra-cluster distances of the nodes from the respective cluster heads. The proposed fitness function to evaluate the strength of the vectors is as follows:

$$fitness = f_1^n + \lambda . f_2^n \tag{3.21}$$

where λ is a regulatory coefficient for the distance of CHs to the BS. The higher the coefficient, the closer the CHs to the base station; f_1^n & f_2^n are the sum of the squared distance between the member nodes and the closest head nodes and sum of squared distances between the cluster heads and the base station respectively.

$$f_1^n = \sum_{i=1}^{k} \sum_{\forall\, Node_j \,\in\, Cluster_i} distance\left(Node_j CH_i\right)$$

$$f_2^n = \sum_{i=0}^{k} distance\left(BS, CH_i\right)$$

As stated above,

$$distance\left(Node_j CH_i\right) = \min_{\forall\, i=1,2,3,\ldots,k} \left\{distance\left(Node_j CH_i\right)\right\}$$

[23] outperforms [11] by minimizing the distance between the cluster heads and the base station via suitable drafting of the fitness function.

Vimalarani C. et al. proposed an enhanced PSO-based clustering energy optimization algorithm for wireless sensor network, EPSO-CEO [24]. The scheme focuses on selection of the cluster heads optimally by applying the steps of particle swarm optimization as per equations (3) & (4). However, the key to optimally formed clusters lies in the proposed fitness function, which not only optimizes the average distances between the cluster members and their respective heads but also brings the residual energies of the member nodes and cluster heads along with the cluster sizes under consideration as follows:

$$fitness = \eta_1 . \frac{\sum_{i=1}^{|Cluster|} distance\ \left(CH_{Cluster}, Node_i\right)}{|Cluster|} + \eta_2 . \frac{\sum_{i=1}^{|Cluster|} RE_i}{RE_{CH}} + \left(1 - \eta_1 - \eta_2\right) . \frac{1}{|Cluster|} \quad (3.22)$$

where, $|Cluster|$ denotes the number of nodes in the cluster; RE_i & RE_{CH} refers to the residual energy of the ith node and cluster head respectively.

In addition to the aforementioned ones, several other contributions have also been made by the researchers, viz. [25–27] in order to address variety of issues in the wireless sensor networks through the use of particle swarm optimization.

3.5 Open Issues and Future Research Directions

This section discusses the main challenges and issues along with the future research directions in order to facilitate readers' further exploration of the process of clustering in WSN.

3.5.1 Challenges and Issues

3.5.1.1 Mobility of the Nodes

Since the inception of the wireless sensor network, mobility has always been considered an important challenge to address. Within the wireless sensor network, there are basically two types of mobility – node mobility and sink mobility. Node mobility refers to the movement of the sensor nodes in which they can change their locations as per the need of application; sink mobility usually indicates autonomous movement of sink node(s), especially in search of data directly from the node deployed in the network.

Be it node mobility or sink mobility, it has to be dealt with carefully as the node/sink movement can introduce serious topological changes in the network.

3.5.1.2 Heterogeneity of the Nodes

Nowadays, wireless sensor networks' deployment requires combinations of heterogeneous sensors for the intended applications, viz. habitat monitoring, environmental monitoring, etc. Here, heterogeneity referring to the nodes with different ability and functionality may present variety of technical issues.

3.5.1.3 Load Distribution

Sensors being power-constrained nodes may suffer with severe problems of energy depletion after which the nodes might fail to participate in the further network operations. In order to minimize the event of early death of nodes, network load must be distributed evenly among the nodes.

3.5.1.4 Scalability

Scalability is a fundamental challenge of clustering. It is necessary to assure that the clustering solution being proposed must be scalable in nature.

3.5.1.5 Energy-Efficient Intra- and Inter-Cluster Communication

As explained already intra-cluster communication refers to the data exchange between the members and head node of a cluster; inter-cluster communication indicates the communication among the heads of clusters. The cluster formation process must deal with the minimization of the aforesaid cost appropriately.

3.5.1.6 Cluster Formation

There are some cluster-specific parameters, viz. cluster count, cluster size, etc. that are crucial for the success of the scheme.

3.5.1.7 Cluster Head Selection

It can be intuited from section 4 that adequate consideration must be given for all the parameters, viz. nodes' proximity, nodes' degree, and nodes' residual energy, etc., affecting the performance of the nodes.

3.5.1.8 Coverage and Connectivity

Based on how the sensors are covering the events in the sensing field, coverage in WSN can be majorly of three types – area coverage, target coverage, and barrier coverage. Area coverage refers to monitoring of the area of interest; target coverage indicates monitoring of the set of specified targets instead of area; and barrier coverage is concerned with the monitoring of a boundary especially for the purpose of intrusion detection. Similarly, connectivity refers to the extent to which the nodes in the field are able to communicate. Adequate consideration for both of the aforementioned parameters is quite important for having an efficient network solution as the applications might require network configurations with different degrees of connectivity and coverage.

3.5.2 Future Research Directions

- Metaheuristic techniques can be called on to effectively solve critical problems of converting the present simulation; especially attention to load-balancing can be focused upon in more and more depth as it controls the performance of the clustering process the most.
- The idea of having multiple mobile sinks can be investigated in order to improve energy efficiency while reducing the energy holes in the network.
- The frequency of cluster heads' rotation can be further optimized to improve the overall network lifetime.
- Security aspects can be also addressed both at the node level and the agent level to further improve the routing schemes.
- Optimization techniques can be integrated with the machine-learning techniques to further improve the clustering process in order to have extended network performance.
- A suitable mixture of different metaheuristic schemes, viz. genetic algorithm, differential evolution, swarm intelligence techniques, and teaching-learning-based optimization etc. can be attempted to improve the network lifetime and experience.
- Adequate consideration of the various parameters such as nodes' proximity, nodes' degree, nodes' location, nodes' residual energy, and nodes' mobility, etc. can be provided towards optimizing the overall clustering process in determination of quality clusters.
- The impact of the physical and data link layers over the clustering process can be investigated and adjusted through the optimization techniques.

3.6 Conclusion

Treating clustering of the nodes as the most significant tool in achieving energy-efficiency and scalability in the wireless sensor networks, the present review emphasizes the clustering process and some metaheuristic clustering approaches especially based on the differential evolution and particle swarm optimization. In addition to this, the present work discusses the clustering taxonomy in detail and surveys a variety of research articles to provide readers with in-depth working and reliable information.

From the above discussion, it can be easily concluded that due to its simplicity, fewer required schematic parameters, and faster convergence, differential evolution has been adapted as the more popular metaheuristic technique for the clustering process in wireless sensor networks. Also, if designed appropriately by considering the suitable set of parameters, viz. node-specific information, nodes' proximity, and nodes' distribution in the sensing field, etc., the fitness function might play a key role in the obtainment of energy-efficient balanced clusters. The challenges and the open research issues along with the future research directions have also been listed extensively, encouraging the readers to explore and discover the new heights in the domain.

References

1. I. F. Akyildiz, W. Su, Y. Sankarasubramaniam, and E. Cayirci, A survey on sensor networks, IEEE Communication Magazine 40 (8), 2002, pp. 393–422, DOI: 10.1109/MCOM.2002.1024422
2. A., Dumka, S. K., Chaurasiya, A., Biswas, and H. L., Mandoria, A Complete Guide to Wireless Sensor Networks: From Inception to Current Trends, 1st Edition; CRC Press, Boca Raton, Florida, USA, 2019.
3. S. Ghiasi, A. Srivastava, X. Yang, and M. Sarrafzadeh, Optimal energy aware clustering in sensor networks, Sensors, 2 (7), 2002, pp. 258–269, DOI: 10.3390/s20700258
4. R. Storn and K. Price, Differential evolution – a simple and efficient heuristic for global optimization over continuous spaces, Journal of Global Optimization, 11 (4), 1997, pp. 341–359.
5. J. Kennedy and R. Eberhart, Particle Swarm Optimization, in Proceedings of ICNN'95-International Conference on Neural Networks, Australia, vol. 4, 1995, pp. 1942–1948.
6. R. V. Rao, V. J. Savsani, and D. P. Vakharia, Teaching–learning-based optimization: a novel method for constrained mechanical design optimization problems, Computer-Aided Design, 43 (3), 2011, pp. 303–315.
7. S. K., Chaurasiya, A., Biswas, and P. K., Bandopadhyay, International Conference on VLSI, Microwave and Wireless Technologies 2021 (ICVMWT-2021), March 20–21, 2021.
8. S. K. Chaurasiya, T. Pal, and S. D. Bit, An Enhanced Energy-Efficient Protocol with Static Clustering for WSN, in Proceedings IEEE Xplore, International Conference of Information Networking (ICOIN), Kuala Lumpur, Malaysia, March 2011, pp. 58–63, DOI: 10.1109/ICOIN.2011.5723134
9. S. K. Chaurasiya, J. Sen, S. Chatterjee, and S. D. Bit, EBLEC: An Energy-Balanced Lifetime Enhancing Clustering for WSN, in Proceeding IEEE Xplore, 14th International Conference on Advanced Communication & Technology-2012, PyeongChang, Korea (South), Feb. 2012, pp. 189–194, INSPEC Accession Number: 12656578.
10. W. R. Heinzelman, A. Chandrakasan, and H. Balkrishnan, Energy-Efficient Communication Protocol for Wireless Microsensor Networks, in Proceedings of 33rd Hawaii International Conference on System Science, Vol. 2, Jan. 2000, pp.1–10, DOI: 10.1109/HICSS.2000.926982
11. W. R. Heinzelman, A. Chandrakasan, and H. Balkrishnan, An application -Specific Protocol Architecture for Wireless Microsensor Networks, IEEE Transactions on Wireless Communications, 1 (4), Oct. 2002, pp. 660–670, DOI: 10.1109/TWC.2002.804190
12. K. Pratyay and K. J. Prasanta, A novel differential evolution based clustering algorithm for wireless sensor networks. Applied Soft Computing, 25, 2014, pp. 414–425.
13. X. Li et al., A differential evolution based routing algorithm for environmental monitoring wireless sensor networks, Sensors, 10 (6), 2010, pp. 5425–5442.
14. C. P. Low, C. Fang, J. M. Ng, and Y. H. Ang, Efficient load-balanced clustering algorithms for wireless sensor networks, Computer Communications, 31 (4), 2008, pp. 750–759.
15. G. Gupta and M. Younis, Load-balanced clustering of wireless sensor networks, in: IEEE International Conference on Communications, ICC03, vol. 3, IEEE, 2003, pp. 1848–1852.
16. K. Pratyay and P. K. Jana, Energy efficient load balanced clustering algorithm for wireless sensor networks, Procedia Technology, 6, 2012, pp. 771–777.
17. S. Potthuri, T. Shankar, and A. Rajesh, Lifetime Improvement in Wireless Sensor Networks using HYbrid Differential Evolution and Simulated Anneailing (DESA), Ain Shams Engineering Journal, (9), 2018, pp. 655–663.

18. J. Brest et al., Self-adapting control parameters in differential evoultion: a comparative study on numerical benchmark problems, IEEE Trans Evol Comput 10 (6), 2006, pp. 646–657.
19. M. Ghahramani and A. Laakdashti, Efficient energy consumption in wireless sensor networks using an improved differential evolution algorithm, 10th International Conference on Computer and Knowledge Engineering (ICCKE), 2020, pp. 18–23, DOI: 10.1109/IC-CKE50421.2020.9303713
20. M., Kaddi, Z., Khalili, and M., Bouchra, A differential evolution based clustering and routing protocol for WSN, 2020 International Conference on Mathematics and Information Technology, Adrar, Ageria, Feb. 18–19, 2020, pp. 190–195.
21. S. Lindsey and C. S. Raghavendra, PEGASIS: Power-efficient gathering in sensor information systems [C], in Proceeding of the IEEE Aerospace Conference, IEEE Aerospace and Electronic Systems Society, 2002, pp. 1125–1130.
22. N. M. A., Latiff, C. C., Tsemenidis, and B. S., Sheriff, Energy-aware clustering for wireless sensor networks using particle swarm optimization, Proceedings of the 18th Annual IEEE International Symposium on Personal, Indoor and Mobile Radio Communications, pp. 1–5 (2007)
23. K. Bennani and D. El Ghanami, Particle swarm optimization based clustering in wireless sensor networks: The effectiveness of distance altering, 2012 IEEE International Conference on Complex Systems (ICCS), 2012, pp. 1–4, DOI: 10.1109/ICoCS.2012.6458564
24. C., Vimalarani, R., Subramanian, and S. N., Sivanandam, An Enhanced PSO-Based Clustering Energy Optimization Algorithm for Wireless Sensor Networks, The Scientific World Journal, Vol. 2016, DOI: 10.1155/2016/8658760
25. B. M., Sahoo, T., Amgoth, and H. M., Pandey, Particle swarm optimization based energy efficient clustering and sink mobility in heterogeneous wireless sensor network, Ad Hoc Networks 106 (2020) 102237, DOI: 10.1016/j.adhoc.2020.102237
26. A., Hassan, A., Anter, and M., Kayed, A robust clustering approach for extending the lifetime of wireless sensor networks in an optimized manner with a novel fitness function, Sustainable Computing: Informatics and Systems 30 (2021) 100482, DOI: 10.1016/j.suscom.2020.100482
27. S., Phoemphon, C., So-In, and N., Leelathakul, Improved distance estimation with node selection localization and particle swarm optimization for obstacle-aware wireless sensor networks, Expert Systems with Applications 175 (2021) 114773, DOI: 10.1016/j.eswa.2021.114773

4

Protocol Development Using Computational Intelligence in Wireless Sensor Networks

Dhirendra Kumar Sharma

CONTENTS

4.1 Introduction

A Wireless Sensor Network (WSN) is a wireless network in which sensor nodes are allowed to establish path via intermediate nodes to other nodes (destination) that are within their transmission range. This chapter presents the concept of WSN and shows its practical applications in today's world.

Moreover, wireless multi-hop sensor networks have been the focus of great attention of the research community for the past few years. Wireless technology is gradually seeping into civil and battlefield surveillance and will become ubiquitous in the near future. It has already shown its applications in e-commerce, business, vehicular services, shopping malls, battlefields, search-and-rescue operations, conferences, virtual classrooms, IoT, etc. Although providing various services, WSN networks are prone to inherent technical challenges, which in turn motivate my research work as described in this chapter along with prominent contributions being made. The characteristic features like self-adaptation and self-organization save lots of resources of the network. The research methodology that guided the development of this research work is also explained, so that appropriate solutions can be obtained for deployment in real networks. For the past couple of years, this general concept of multi-hop communications in wireless networks has developed a number of specializations of different kinds of networks (WSN, Wireless Mesh Network [WMN], and Vehicular Adhoc Network [VANET] [1, 2]). These specializations differ on the basis of the mobility of the nodes, their processing power,

DOI: 10.1201/9781003102397-4

and energy efficiency. Although all of them follow the same basic principles of wireless multi-hop communications, they also have noticeable distinctive features that make it impractical to achieve a panacea. By examining the literature, the routing protocols are mainly proactive, reactive, and hybrid. The tremendous success of ad hoc networking and WSN are due to the surplus services that WSN offers through all the variants discussed above. Nevertheless, before providing a service, there are numerous technical challenges that must be met. For allowing applications to send and receive data packets, the ad hoc network should establish the best possible path (according to the chosen metric) between sender and receiver nodes. On the basis of the number of destination nodes of the data flow, we can identify various types of communications. For instance, unicast communications are intended for a single destination. If multiple destination nodes exist, then we are referring to a multicast communication. However, when the data must be delivered to all the nodes in the network, the communication is said to be of broadcasting type. If only those nodes that are, say, "n" hops away from the source are the destinations, we are talking about topology-constrained broadcasting. Finally, if the set of destinations comprises all nodes existing in only a given geographic area, this type of communication is called geo-casting or geo-broadcasting. To employ routing protocols in ad hoc networks successfully, various other technical issues must also be handle, such as route stability and its maintenance over network. As a matter of fact, designing and developing protocols for wireless sensor networks is a challenging task. Here, our focus is on the analysis and design of efficient, adaptive, and scalable routing protocols for WSN. The next section presents the objectives of the work performed in the context of this chapter [1, 3].

4.1.1 Objectives

The main objective of this chapter is explore WSN routing and MAC protocols that are appropriate for challenging environments. Given the inherent characteristics of these networks, the solutions must be adaptable to dynamic topology, efficient with the bandwidth usage, and scalable and energy efficient when various network parameters are concerned. Moreover, the focus is also on realistic approaches having relevance in real-life deployments. This means that the protocols should not be designed merely on the basis of generic assumptions, which could lead to incorrect conclusions. Network Simulator (ns2.34) is the tool used to determine that the developed algorithms are implementable in real networks. For validation in ad hoc networks, a specific scenario should also be mentioned for which the routing protocol has been designed [2].

Meeting these objectives includes providing the perception of the realities of ad hoc networking. In addition, the proposed solutions should be compared against corresponding solutions found in the literature. The proposed solutions must offer better performance with respect to others in order to be able to contribute something to the research community. To summarize, the main goal of this chapter is to improve the knowledge of wireless sensor networking by providing solutions that can help in developing new features for WSNs. The next section gives the details of the scope of methodology.

4.1.2 Scope

This chapter focuses on the uses of CI for routing and MAC protocol development, which improves overall performance of networks. Contributory works of this thesis are related to battlefield and rescue operation (examples of dense network) over static and dynamic topology. For such type of operations, we require adequate node density and overhearing capacity. Our proposed works are simulated in Network Simulator (ns2.34) and our work focuses only on the core architecture of simulation. We have used a simple hello process to exchange the important parameters to control the routing and maintenance process. We are assuming hello packets for exchange of priority without adjusting actual size of hello message. Mainly our research works related to following domains:

- Use of CI for routing and network maintenance to improve significant performance of WSN.
- Utilization of MAC protocols techniques for performance improvement.

4.2 Related Work

In this section we discuss the existing routing protocol and MAC protocols. Here, mainly we emphasize an ant routing algorithm. WSN routing protocols are mainly static and dynamic. In the case of static routes, metrics remain the same for data transfer. To provide the support in mobile environment, dynamic approach updates the route metrics information [4, 5].

4.2.1 Proactive Routing Protocol

Proactive routing protocol is classified as distance vector routing and link-state routing; in a distance vector routing algorithm, each mobile has routing information in the routing table. This routing table has an entry about destination node, hop count, and next hop information. In distance vector routing, data packets follow the routing table entry. If an entry of a mobile node does not exist then it sends the data to the adjacent node and data packets are transmitted. In DSDV and WRP distance vector routing protocols, DSDV is the first routing protocol and it is based on the Bellman-Ford algorithm. Using cryptographic primitives enhance the security for routing protocols, Secure Efficient Ad hoc Distance (SEAD) vector routing protocol follows the concept of efficient one-way hashing technique that provides the authentication-based service for updating the entry and improving the performance of the routing protocol [5–8].

Usually, distance vector routing creates a lot of route overheads for maintaining the routing table entry so multi-point relays (MPRs) are used to diminish the routing overheads in the network. Optimized linked state routing (OLSR) protocol uses the MPR concept. Using MPR efficiently disseminates link state updates across the network; it allows mobile nodes to generate the link state update message. Basically OLSR uses only periodic updates for link-state dissemination, due to this total overhead determined by the product of number of nodes generating the updates; it gives the best performance for large networks [9, 10]. Recently, an experimental result of OLSR in a wireless grid lab showed the multi-hop characteristics and explained that a grid structure does yield a worst-case complexity problem in terms of availability of alternate routes [11].

In large-scale span area, WSN resources are deployed as sensor nodes to form a region of interest from where data accumulated at base station (BS). Here one major problem comes in terms of long-term monitoring data transmission to BS. An energy-efficient proactive routing approach where node degree affects the lifetime. In order to handle this situation a collaborative distributed antenna routing protocol is proposed in which the physical parameter of a node as antenna is considered for optimum energy transfer [12].

Power always a major issue in WSN; when we are using its application for underwater communication it becomes a more serious concern. We know that replacing a battery is a very tedious task in underwater networks. It can be resolved by controlling overhead, less failure system. Here cluster-based, proactive energy efficient routing helps to solve the problem [13, 14].

4.2.2 Reactive Routing Protocol

In WSN a gradual updating process of routing table entry creates chaos in the database entry of routing table, so a reactive routing protocol has been developed to curtail this problem. Basically it is on-demand based routing; when a route is required it starts to establish the connection. Reactive routing protocols follow two phases: route discovery and route maintenance. The first phase is used to establish the connection, and another phase is required when the connection is destroyed. Recently, a survey of multi-path routing has shown that the source node uses the multiple disjoint paths for routing data packets [15, 16]. In this scheme it limits the rebroadcast scheme and, during the connection establishment process, avoids routing loops in the network.

AODV uses the intermediate node for data transfer, and during the route maintenance process a route error message is sent to the source node and the source node starts reroute discovery. Now backup route concepts are used in which route error messages are sent on alternate paths. Operation of alternate paths works efficiently if an adjacent node exists in the network at a broken link. Existing routing protocol works efficiently and gives the best performance for a static network; even for a dynamic network its reliability decreases rapidly. Suppose an adjacent node does not exist in the network; then route failure is caused automatically and packet loss occurs [17].

An energy efficient DSR routing protocol follows the route-salvaging process; it selects the path from the source node whenever route failure occurs; it erases the entry of the broken link and selects the alternate path from its routing table. If an alternate path does not exist, then it starts the rerouting discovery process. Use of greed-based backup routing gives high stability in mobile ad hoc networking (MANET); basically it uses the link lifetime and route length as a weight metric [18]. It uses the greedy forward mechanism to establish the route and local backup path and considers the link lifetime for route repair mechanism [18].

4.2.3 Hybrid Routing Protocol

Energy consumption is a serious issue in wireless sensor networks. Proactive and reactive routing protocols have discrete properties, but hybrid routing merges both properties and reduces the demerits of both routing protocols. In the case of hybrid routing, the optimal routing strategy depends on the underlying network topology, variable data rate, and heavy traffic pattern, which varies dynamically. In the paper author innovates the sharp hybrid adaptive routing protocol (SHARP), which automatically finds the balance point between proactive and reactive routing by adjusting the degree to which route information is propagated proactively versus the degree to which it needs to be discovered reactively. SHARP enables each node to use a different application-specific performance metrics to control the adaptation of the routing layer. It describes application-specific protocols built on top of SHARP for minimizing packet overhead, bounding loss rate, and controlling jitter. Various heterogeneous routing scheme shown in Figure 4.1 [19, 20].

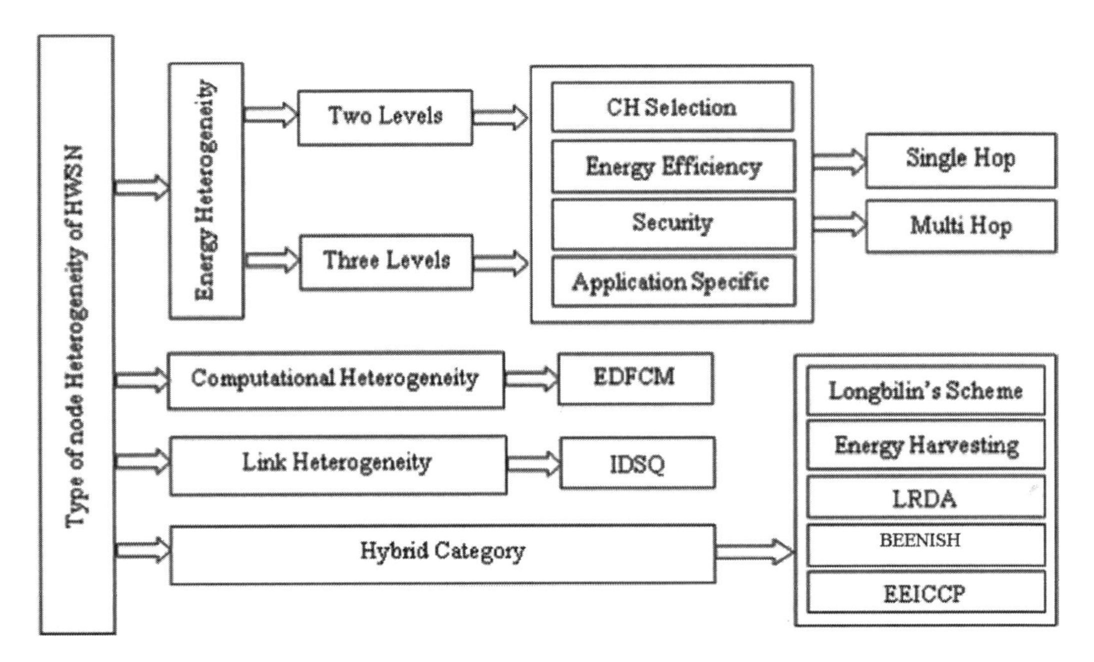

FIGURE 4.1
Taxonomy of various heterogeneous routing schemes [20].

Generally, in WSN, two-zone routing protocol (TZRP) is a nontrivial expansion of ZRP. In contrast with the original ZRP, in which a single zone serves a dual purpose, TZRP aims to decouple the protocol's ability to adapt to traffic characteristics from its ability to adapt mobility. In support of this goal, in TZRP each node maintains two zones: a crisp zone and a fuzzy zone. By adjusting the sizes of these two zones independently, a lower total routing control overhead can be achieved. The working of TZRP is based on a MANET routing framework that can balance the trade-offs between various routing control overheads more effectively than ZRP in a wide range of network conditions [21].

In WSN, core extraction distributed ad hoc routing (CEDAR) uses a technique that integrates the routing and quality of service. CEDAR uses the core broadcast mechanism to transmit the data packets throughout the networks. In the network, the core node maintains its adjacent local topology information. It employs an efficient link state propagation mechanism in which information regarding the presence of high bandwidth and stable link propagates through several links. During the route maintenance process, it repairs the broken route locally; it uses a route error message, and after receiving this message the source node starts recomputation of a route from itself to the destination [22].

A scalable proactive protocol routing (S-OLSR) is used to increase the coverage area of routing for sensor nodes in a geographical span area [23].

4.2.4 Ant Colony Optimization Routing Protocol

A survey of ant routing shows the properties of ants for routing; basically it follows the foraging process; in this process ants do not know about their path, and they randomly search the path. They use a special type of chemical for communication; it is called a pheromone. After searching for food an ant creates the shortest path and all ants follow that and maintain the intensity of pheromone. Usually, an ACO routing algorithm uses control

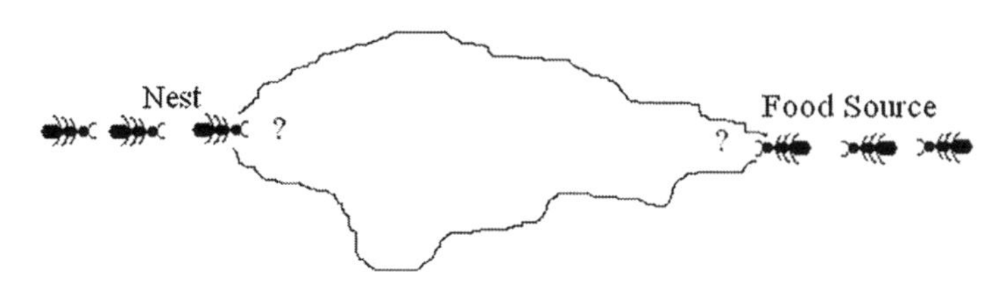

FIGURE 4.2
Foraging process.

packets to provide the acquisition of routing information for routing, and that is known as ants. Usually, a large amount of routing information exchange creates congestion, which is responsible for packets drops and jitter; the whole process is similar to the traffic pattern of ants on their path. Recently, ant routing has been used for ad hoc networks and it is known as AntHocNet; it uses the forward ant and backward ant for establishment of path, as shown in Figures 4.2 and 4.3. In AntHocNet, the source uses forward ants (F-ANT) to search the destination and backward ants (B-ANT) are used for confirmation of the path. After establishing a connection, route source node always chooses the shortest path and sends the data packets stochastically, but these cases are adopted for multipath routing. Another GPS/Ant-based routing protocol has been proposed for MANET; basically, the use of a GPS-based system provides the exact location of a mobile node. The GPS-based approach assumes that some mobile node positions are fixed and it updates the routing table periodically with its dynamic neighbor node [24–26].

In a paper an author proposed distributed ant routing (DAR). This approach explains the self-organization of datagrams in MANET; locating and storing are the main functions in this routing. Datagrams are routed deterministically, which is based on the maximum probability at each intermediate node from source node to the destination node. This paper minimizes the complexity and gives the optimal solution for critical connectivity [27].

Swarm intelligence is a congestion-aware protocol; due to a self-organization mechanism it has the ability to measure the slow deterioration of a link and restore a new link according to the requirement of the network. ANSI uses the local route repair process and sends the data packets after route repair, but here latency increases when it applies this process at the starting or source node. In our proposal we reduce the latency; first we find the position of link break and then start the local route repair process [28].

In a review paper the author describes three factors that moderate the stagnation of pheromone intensity: aging, delaying, noise. These three factors reduce the flow rates of ants from a congested node to its adjacent in the network. Suppose that if flow of ants reduces due to delayed ants from node i to $i + 1$ decreases the intensity of pheromone another thing is that time delay large then the pheromone intensity becomes zero and path is destroyed [29].

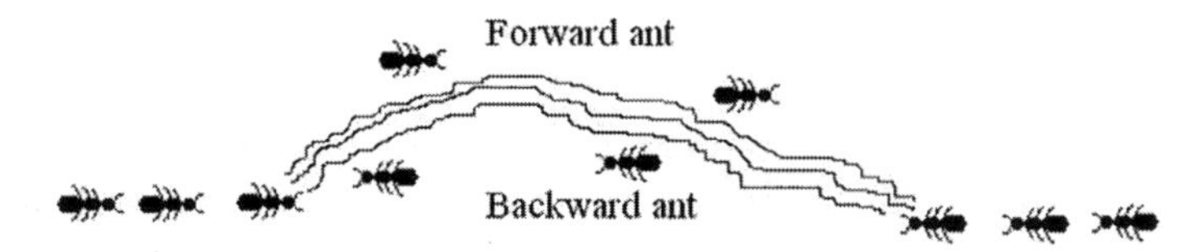

FIGURE 4.3
Traffic pattern of ants.

In WSN, the use of a large number of control packets is the cause of congestion and traffic; it is reduced by a flow-control mechanism; the use of a rate-based mechanism reduces the traffic. A flow-control mechanism is described as a window-based flow control and predictive control mechanism. Window-based flow control is a kind of closed-loop control in which the receiver keeps a buffer of size W data units for a particular session, where W is called the window size. The predictive control mechanism uses time-delayed control systems, backlog balancing flow control, and hop-by-hop rate base congestion control [30].

In ref. [31], the author describes the collective behavior of ants. When any obstacles come on the path, the ants immediately search for a new path. This collective behavior is called stigmergy. While foraging ants use a special type of chemical known as a pheromone; they deposit the pheromone on the surface. Ant-based routing protocol uses an. During this process ants erase the path. In ant colony optimization the most effective and intelligent ants are Argentine ants. In this paper, with the help of ant-based routing protocol, we propose a heuristic approach that is useful during the route renewal and route maintenance process. In Figures 4.2 and 4.3 we describe the ant routing.

Cluster-based routing in WSN also helps to reduce the energy consumption, which is essential for increasing the lifetime of a network. During the selection of cluster nodes, energy is considered to avoid failure [32]. In the case of an intra-cluster communication mobile node (MN), balanced energy is estimated to predict the lifetime of MN communication [33].

A fuzzy logic–based routing method is proposed to resolve the routing issues for WSN and MANET, in which an incremental approach is used for route discovery and route maintenance [34]. In cluster-based WSN a fuzzy logic–based jamming detection algorithm (FLJDA) method is proposed to detect the jamming in downstream path where jamming metrics computed by FLJDA [35].

4.2.5 Medium-Access Control Protocols

We know that the medium access control protocol is used to access the channel for effective transfer of data between mobile nodes. In a WSN, a better channel access mechanism optimizes the performance parameters. Generally, issues arise in a highly dense active network where several mobile nodes attempt to transfer the data. In the case of frame loss backoff mechanisms estimate the waiting time for a re-transmission attempt. Loss of frame may be control (RTS, CTS, ACK) and data type; a backoff mechanism uses control metrics as types of contention window (CW). A contention window has several values: default, minimum, maximum, and threshold. These values estimate the final backoff time [36].

MAC protocols perform several tasks, such as channel allocation; they are used to reduce interference during communication. There are several types of channel allocation, such as static and dynamic. In a static channel allocation scheme, all mobile nodes are allocated with discrete channels for transfer of data frames. In the case of dynamic channels, they are allocated on the basis of transmission node density. Collision avoidance in wireless networks becomes more challenging when all mobile nodes are transmitting simultaneously. The IEEE 802.11 distributed coordinated function (DCF) standard has CSMA/CA and backoff algorithm concepts to cope with such type of issues. A backoff algorithm is used to estimate the waiting time (backoff time; tbackoff) for retransmission of lost or unacknowledged packets. Actually, a backoff algorithm uses the contention resolution algorithm, which became significant for certain network configurations, such as noisy channel and congested network. Besides these facts, a scenario of an increase in hidden nodes causes a high collision probability, which rapidly decrease the network performance. Here, backoff tuning may give excessive attempts for successful transmission of packets, which may lead to increased delay. To illustrate the backoff algorithm functionality we have considered an example in Figure 4.1, which shows the traditional backoff process.

DCF is a basic medium-access control mechanism that uses retransmission for drop packets, this approach becomes ineffective in the case of high traffic and node density. An efficient DCF algorithm shows performance enhancement when mobile nodes are moving. Here, the existing algorithm became less effective in frequent link failure, so CSMA/CA uses its effective strategy for data frame transfer. When a mobile node senses the medium is free, it starts the data transmission, otherwise it keeps on waiting. For successful reception of packets, CSMA/CA protocol uses a positive acknowledgment (ACK) scheme. After a short time interval, the receiving node sends the ACK packet to sender node. In the case of unsuccessful delivery of ACK packet, the sender arranges the retransmission of packets, as shown in Figure 4.4. In a shared wireless medium collisions of packets are reduced by adjustment of contention window for backoff estimation. In this paper, a contention window threshold is used to adjust the backoff time for retransmission of dropped packets. A binary exponential backoff (BEB) algorithm works better in low-density regions where traffic density is low. In addition, when node density increases in the medium, a node increases the backoff slot number; hence the waiting time increases. As we know, major backoff algorithms reduce the contention window value on several factors. Traditional backoff algorithm methods tend to reduce backoff time. In the papers authors developed the MAC protocols that use the BEB algorithm for contention resolution. However, in congested and noisy scenarios the backoff parameter does not work to avoid collisions. Such a scenario increases the collision probability and degrades the channel utilization. In the case of low-level congestion in the system, a small backoff window is considered. When the level of congestion increases, then a small backoff value fails to handle the packet loss. In mobile ad hoc network mobile node free to transmit the data packet at any time for this frequent calling of backoff incurs but in this scenario minimum throughput achieved and increases the end-to-end delay. For a practical scenario, wireless links are unreliable and noisy, so maximum path loss occurs in the channel due to fading, and interference causes bit errors. During the initialization, when the first transmission attempt occurs, a backoff timer takes the minimum contention window (CWmin). For every unsuccessful transmission, the contention window size becomes double until it reaches the maximum backoff window size (CWmax). When the CW is set to CWmax until it is reset, for successful transmission it decreases linearly or multiplicatively or exponentially. ACK frames are sent by the destination when the transmission is successfully received [37].

There are some similarities in wireless personal area between wireless sensor networks. In the IEEE 802.11 wireless local area networks (WLANs) standard, network nodes experiencing collisions on the shared channel need to backoff for a random period of time, which

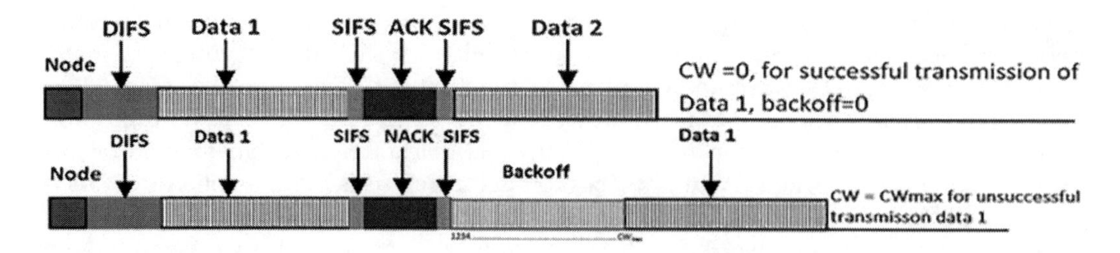

FIGURE 4.4
IEEE 802.11 DCF backoff tuning.

is uniformly selected from the contention window. This contention window is dynamically controlled by the BEB algorithm. The BEB scheme, suffers from a fairness problem that shows low throughput under high traffic load. To eliminate this fairness problem an enhanced version of this algorithm is the enhanced binary exponential backoff algorithm (EBEB) for successful transfer of data packets. In EBEB, a constant counter variable is used to maintain the contention window. The counter variable accounts for several successful attempts. Our proposed method is the modification of EBEB algorithm for improvement of MANET performance. We have assumed that two separate counters, namely C1 and C2, are used to encounter both the successful transmission state and unsuccessful transmission state respectively. In the case of the EBEB algorithm, only the successful transmission state was taken into consideration, which caused ambiguity. The EBEB algorithm works only for successful transmission, but in our proposal, we have checked for unsuccessful transmission by applying a condition to check the delivery of the data frame. In the case of unsuccessful transmission, the C2 counter is initialized to count the number of unsuccessful deliveries of data frames. And before adjustment of the contention window, we compared the last two values of counters C1 and C2. In order to get rid of this ambiguity both the cases are taken into account. The flow chart in Figure 4.4 shows that there are two separate flows for the successful and unsuccessful transmission by the node. The successful transmission is same as that of EBEB algorithm. The difference lies in the case of unsuccessful transmission. Here first the value of unsuccessful counter C2 is compared with the maximum threshold value and then further compared with the value of counter C1. According to the higher value amongst the two, the further flow chart is carried out as shown in Figure 4.5 [38].

4.2.6 Fuzzy-Logic–Based MAC Protocols

In WSN lifetime reduction is a serious concern, so an optimization technique emerged as an effective solution. Several fuzzy-logic–based algorithms were developed for MAC and network layers for increasing the lifetime. A distributed fuzzy-logic–based sink selection algorithm reduces congestion over sink nodes and provides load balancing. There is a reduction in retransmission, which shows that overall the performance of WSN increases in terms of maximum lifetime in a heavy congestion scenario [39, 40].

Another way of improving the lifetime of sink nodes inside a cluster is selecting the optimum number of intermediate cluster heads. Multi-level fuzzy logic concept can be applied for finding the most suitable cluster heads. This way lifetime of cluster increased to gain efficiency [41].

Rechargeable sensor node concepts emerged and address placement of sink nodes, frequency selection, and number of RF-energy transmitters. RF-MAC optimizes energy delivery sensor nodes [42].

In WSN bandwidth utilization is a main concern. Its unregulated use causes degradation in performance parameters. A fuzzy-logic–based contention window optimization improves the performance. Here CW optimizes dynamically by use of a fuzzy-logic controller to analyze the number of collisions [43].

Uncontrolled access of channels causes a fairness issue, which shows overall underutilization of resources. We understand that lower CW values cause this issue. An impulse radio ultra wideband (IR-UWB) method uses the fuzzy-logic–based technique to increase channel utilization for WSN [44].

An increase in the number of tiny devices increases network connectivity and abruptly causes maximum packets drops. To avoid such issues an efficient MAC strategy is required

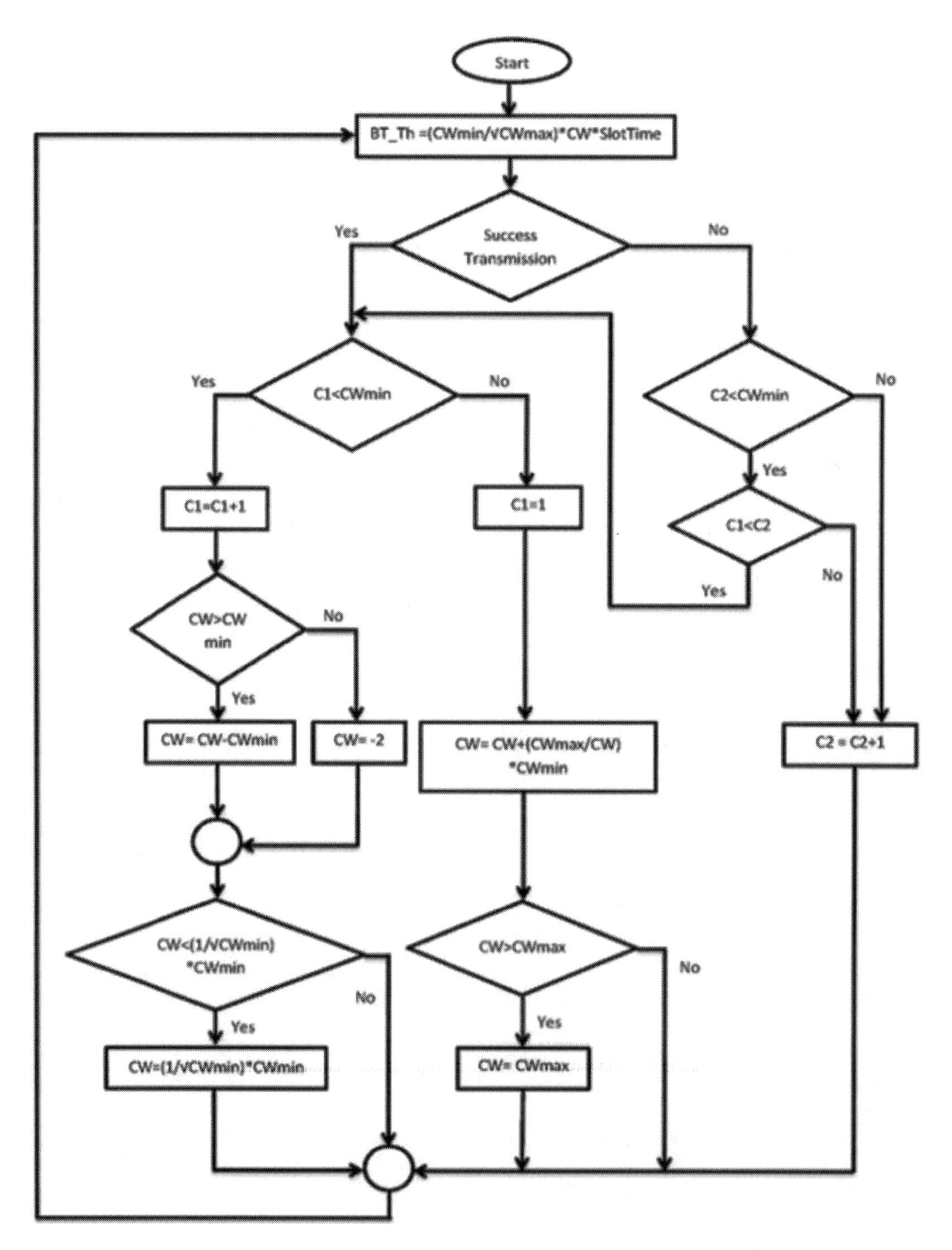

FIGURE 4.5
Adjustment of contention window.

to increase network survivability. A survivability aware channel allocation (SACA) has emerged to counter such network issues [45].

In a recent trend, energy harvesting is very important for long-term communication. Here energy optimization helps to increase the lifetime of a network. A network lifetime is increased by adaptive traffic control improvement by defining priorities for nodes. Nodes assign their priorities on the basis of traffic rate and packet delay. Fuzzy-logic works on queue length and working with a CSMA/CA algorithm [46, 47].

Recently industrial WSN (IWSN) has been using sensors for several types of applications. This requires better WSN network architecture to support IWSN. Fuzzy-logic used to classify the quality channels includes graylist, whitelist, and allowlist later it applied over time slotted channel [48].

In WSN mesh topology all communication devices exchange high amounts of data, which is similar to internet of things (IoT) applications. It can be understood as home automation where multiple sensors are exchanging data with the central coordinator. Here issues arise with the channel access mechanism where multiples sensor nodes are accessing the channel simultaneously. In order to handle such a situation modified CSMA/CA protocols emerged to handle real-time access of data exchange with effective quality of service (QoS) [49].

4.3 Conclusions

In WSN several challenges exist for routing and MAC protocols development. When a route is required, initially a mobile node or sensor nodes invoke the route discovery to find the optimum path. Route discovery works by exchanging information such as next hop information from the neighbor table. Routes are computed in the initial phase of route discovery and later updated by route maintenance. This method helps to make use of routes for a long time where less mobility exists. Nature-inspired routing also confirms selection of the optimum route by computational intelligence, like ant routing. Use of computational intelligence helps to identify optimum paths for efficient data transfer. In order to experience better performance, we need to focus on MAC protocol functioning. Channel allocation issues are effectively improved by MAC protocol functioning. In the case of a packet collision scenario, a better backoff mechanism shows effectiveness on performance parameters. Overall in this work we tried to explore routing and MAC protocols that require CI to attain better performance for wireless sensor networks.

References

1. A. Komathi, M. Pushparani. "Trust performance of AODV, DSR and DSDV in wireless sensor networks," Second International Conference on Current Trends in Engineering and Technology - ICCTET 2014, 2014, pp. 423–425.
2. https://www.isi.edu/nsnam/ns/
3. S. Carson. Mobile Ad hoc Networking (MANET) working group. Internet Draft, www.ietf.org/rfc/rfc2501.txt

4. Mirjana Maksimović, Vladimir Vujović, and Vladimir Milošević. "Fuzzy Logic and Wireless Sensor Networks – A Survey", Journal of Intelligent & Fuzzy Systems, Volume 27, No. 2, 2014, pp. 877–890.
5. Jegan Govindasamy and Samundiswary Punniakody. "A comparative study of reactive, proactive and hybrid routing protocol in wireless sensor network under wormhole attack", Journal of Electrical Systems and Information Technology, Volume 5, No. 3, 2018, pp. 735–744.
6. J. L. Deneubourg, S. Aron, S. Goss and J. M. Pasteels. "The self-organizing exploratory pattern of the Argentine ant", Journal of Insect Behavior, Volume 3, No. 2, pp. 159–168.
7. Mehran Abolhasan, Tadeusz Wysocki, and Eryk Dutkiewicz. "A review of routing protocols for mobile ad hoc networks", Ad Hoc Networks, Volume 2, No. 1, 2004, pp. 1–22.
8. Itu Snigdh and Devashish Gosain. "Analysis of scalability for routing protocols in wireless sensor networks", Optik, Volume 127, No. 5, 2016, pp. 2535–2538.
9. Yih-Chun Hu, David B. Johnson, and Adrian Perrig. "SEAD: secure efficient distance vector routing for mobile wireless ad hoc networks", Ad Hoc Networks Volume 1, No. 1, July 2003, pp. 175–192.
10. T. Clausen and P. Jacquet. Optimized Link State Routing Protocol (OLSR). www.ietf.org/rfc/rfc3626.txt
11. Anelise Munaretto and Mauro Fonseca. Routing and quality of service support for mobile ad hoc networks. Computer Networks, Volume 51, 2007, pp. 3142–3156.
12. Reem E. Mohamed, Walid R. Ghanem, Abeer T. Khalil, Mohamed Elhoseny, Muhammad Sajjad, and Mohamed A. Mohamed. "Energy efficient collaborative proactive routing protocol for Wireless Sensor Network", Computer Networks, Volume 142, 2018, pp. 154–167.
13. Z. A. Khan et al. "Region Aware Proactive Routing Approaches Exploiting Energy Efficient Paths for Void Hole Avoidance in Underwater WSNs," IEEE Access, Volume 7, 2019, pp. 140703–140722.
14. Reem E. Mohamed, Walid R. Ghanem, Abeer T. Khalil, Mohamed Elhoseny, Muhammad Sajjad, and Mohamed A. Mohamed. "Energy efficient collaborative proactive routing protocol for Wireless Sensor Network", Computer Networks, Volume 142, 2018, pp. 154–167.
15. Broch J, Johnson DB, and Maltz D A. The Dynamic Source Routing for Mobile Ad Hoc Networks. IETF (1998), RFC 4728.
16. Mohammed Tarique, Kemal E. Tepe, Sasan Adibi, and Shervin Erfani. Survey of multipath routing protocols for mobile ad hoc networks. Journal of Network and Computer Applications, Volume 32, No. 6, 2009, pp. 1125–1143.
17. Wei Kuang Lai, Sheng-Yu Hsiao, and Yuh-Chung Lin. Adaptive backup routing for ad-hoc networks. Computer Communications, Volume 30, 2007, pp. 453–464.
18. K. Ng and C. Tsimenidis, "Energy-balanced dynamic source routing protocol for wireless sensor network," 2013 IEEE Conference on Wireless Sensor (ICWISE), 2013, pp. 36–41
19. N. Sengottaiyan, R. Somasundaram, and A. Balasubramanie. Hybrid Routing Protocol for Wireless Sensor Network. In: Das V.V. et al. (eds) Information Processing and Management. BAIP 2010. Communications in Computer and Information Science, Volume 70, 2010, pp. 188–193.
20. Sudeep Tanwar, Neeraj Kumar, and Joel J.P.C. Rodrigues. "A systematic review on heterogeneous routing protocols for wireless sensor network", Journal of Network and Computer Applications, Volume 53, 2015, pp. 39–56.
21. Muni Venkateswarlu Kumaramangalam, Kandasamy Adiyapatham, and Chandrasekaran Kandasamy. "Zone-Based Routing Protocol for Wireless Sensor Networks", International Scholarly Research Notices, Volume 10, No. 12, 2014, pp. 1506–10523.
22. Raghupathy Sivakumar, Prasun Sinha and Vaduvur Bharghavan. Core Extraction Distributed Ad hoc Routing (CEDAR) Specification. Internet Draft: draft-ietf-manet-cedar-spec-00.txt.
23. Sukhleen Jaggi and Er. Vikas Wasson. "Enhanced OLSR routing protocol using link-break prediction mechanism for WSN." Industrial Engineering and Management Systems, Korean Institute of Industrial Engineers, Volume 15, No. 3, September 2016, pp. 259–267.

24. Luis Cobo, Alejandro Quintero, and Samuel Pierre. "Ant-based routing for wireless multimedia sensor networks using multiple QoS metrics", Computer Networks, Volume 54, No. 17, 2010, pp. 2991–3010.
25. Marco Dorigo, Eric Bonabeau and Guy Theraulaz. Ant algorithms and stigmergy. Future Generation Computer Systems, Volume 16, 2000, pp. 851–871.
26. Daniel Camara and Antonio A. F. Loureiro. GPS/Ant-Like Routing in Ad Hoc Networks. Telecommunication Systems, Volume 18, No. 1–3, 2001, pp. 85–100.
27. Sundaram Rajagopalan and Chien-Chung Shen. ANSI: A swarm intelligence-based unicast routing protocol for hybrid ad hoc networks. Journal of Systems Architecture, Volume 52, 2006, pp. 485–504.
28. Laura Rosati, Matteo Berioli, and Gianluca Reali. On ant routing algorithms in ad hoc networks with critical connectivity. Ad Hoc Networks, Volume 6, 2008, pp. 827–859.
29. Kwang Mong Sim and Weng Hong Sun. Ant Colony Optimization for Routing and Load-Balancing: Survey and New Directions. IEEE Transactions on Systems, MAN, and Cybernetics-Part A: Systems and Humans, Volume 33, No. 5, September 2003, pp. 560–572.
30. H. Jonathan Chao and Xiaolei Guo. Quality of Service Control in High-Speed Networks. John Wiley & Sons, Inc, ISBN 0-471-22439-1, 2002, pp. 238–240.
31. Li Ting, Tang Rui-bo, and JI Hong. Status adaptive routing with delayed rebroadcast scheme in AODV-based MANETs. The Journal of China Universities of Posts and Telecommunications, Volume 15, No. 3, September 2008, pp. 82–86.
32. Duo Peng and Qiuyu Zhang. "An energy efficient cluster-routing protocol for wireless sensor networks," 2010 International Conference on Computer Design and Applications, 2010, pp. V2-530–V2-533.
33. N. Gharaei, K. A. Bakar, S. Z. M. Hashim, and A. H. Pourasl. "Energy-Efficient Intra-Cluster Routing Algorithm to Enhance the Coverage Time of Wireless Sensor Networks," IEEE Sensors Journal, Volume 19, No. 12, pp. 4501–4508.
34. A.M. Ortiz, F. Royo, T. Olivares, et al. Fuzzy-logic based routing for dense wireless sensor networks. Telecommunication Systems, Volume 52, 2013, pp. 2687–2697.
35. K. P. Vijayakumar, P. Ganeshkumar, M. Anandaraj, K. Selvaraj, and P. Sivakumar. Fuzzy logic–based jamming detection algorithm for cluster-based wireless sensor network. International Journal of Communication Systems, Volume 31, 2018, p. e3567.
36. P. R. Rothe and J. P. Rothe "Medium Access Control Protocols for Wireless Sensor Networks", Handbook of Wireless Sensor Networks: Issues and Challenges in Current Scenario's, Volume 1132, 2020, pp. 35–51.
37. M. Al-Hubaishi, T. Alahdal, R. Alsaqour, A. Berqia, M. Abdelhaq, and O Alsaqour. Enhanced binary exponential backoff algorithm for fair channel access in the ieee 802.11 medium access control protocol, International Journal of Communication Systems, Volume 27, 2014, pp. 4166–4184.
38. Dhirendra Kumar Sharma and Rohit Srivastava. "An impact of efficient backoff algorithm in MANETs", VLSI, Microwave and Wireless Technologies – Proceedings of ICVMWT 2021, MMMUT, Gorakhpur. Springer, March 2021, pp. 20–21.
39. Ashutosh Kumar Singh, N. Purohit, and S. Varma. Fuzzy logic based clustering in wireless sensor networks: a survey, International Journal of Electronics, 2013, Volume 100, No. 1, pp. 126–141.
40. M. Masdari and F. Naghiloo. Fuzzy Logic-Based Sink Selection and Load Balancing in Multi-Sink Wireless Sensor Networks. Wireless Personal Communications, Volume 97, 2017, pp. 2713–2739.
41. Ashutosh Kumar Singh, N. Purohit, and S. Varma. "Fuzzy logic based clustering in wireless sensor networks: a survey", International Journal of Electronics, Volume 100, No. 1, 2013, pp. 126–141.
42. M. Y. Naderi, P. Nintanavongsa, and K. R. Chowdhury. "RF-MAC: A Medium Access Control Protocol for Re-Chargeable Sensor Networks Powered by Wireless Energy Harvesting," IEEE Transactions on Wireless Communications, Volume 13, No. 7, July 2014, pp. 3926–3937.

43. M. Collotta, A. L. Cascio, G. Pau, and G. Scatá. "A fuzzy controller to improve CSMA/CA performance in IEEE 802.15.4 industrial wireless sensor networks," 2013 IEEE 18th Conference on Emerging Technologies & Factory Automation (ETFA), 2013, pp. 1–4.
44. Darif Anouar, Aboutajdine Driss, and Saadane Rachid. "A New Backoff Algorithm for a Low Power Consumption MAC Protocol in IR-UWB Based WSN", Journal of High Speed Networks, Volume 24, No. 3, 2018, pp. 175–185.
45. Manu Elappila, Suchismita Chinara, and Dayal Ramakrushna Parhi. "Survivability Aware Channel Allocation in WSN for IoT applications", Pervasive and Mobile Computing, Volume 61, 2020.
46. Imen Bouazzi, Monji Zaidi, Mohammed Usman, and Mohammed Zubair M. Shamim. "A new medium access control mechanism for energy optimization in WSN: traffic control and data priority scheme", EURASIP Journal on Wireless Communications and Networking, Volume 2021, No. 42, 2021.
47. Ramakrishnan Sabitha and Thangavelu Thyagarajan, "Fuzzy logic-based transmission power control algorithm for energy efficient MAC protocol in wireless sensor networks", International Journal of Communication Networks and Distributed Systems, Volume 9, No. 3–4, 2012, pp. 247–265.
48. D. V. Queiroz, R. D. Gomes, I. E. Fonseca et al. Channel assignment in TSCH-based wireless sensor networks using fuzzy logic. Journal of Ambient Intelligence and Humanized Computing, 2021.
49. Achour Achroufene, Mourad Chelik, and Nassima Bouadem. "Modified CSMA/CA protocol for real-time data fusion applications based on clustered WSN", Computer Networks, Volume 196, 2021, p. 108243.

5

Intelligent IoT for Precision Agriculture

Hitesh Kumar Sharma

CONTENTS

5.1 Introduction

The IoT (or wireless communication sensors) is a collection of integrated provisioning units, known as sensors, that communicate over wireless networks. The collection of restrictions connected to the general environment, such as temperature, pressure, or the existence of items, is its basic limit. They usually have a slew of sensors that communicate through radio to provide information and make getting ready more enjoyable. An automated irrigation method distributes water to crops in the field by spraying like a natural rainfall. Installing a smart irrigation system saves time and ensures judicious usage of water. An IoT-enabled on-farm device effectively passes on inputs. Artificial intelligence (AI) is used to create an individual plan for each crop to track the yield production and helps to generate most suitable plan for next crop [1, 2]. Data will provide secured pieces of information on how to best regulate food creation.

A much more promising option for agriculture to deal with issues such as crop change, water practicality, and smart water system management is to continue to improve the growing collaboration by designing smart buildings and IoT devices to collect and manage important agricultural data in a compelling manner (Figure 5.1 (a and b)). Farm information can be gathered in a variety of ways, including the human connection on mobile applications or paper, mobile platforms, or customized

DOI: 10.1201/9781003102397-5

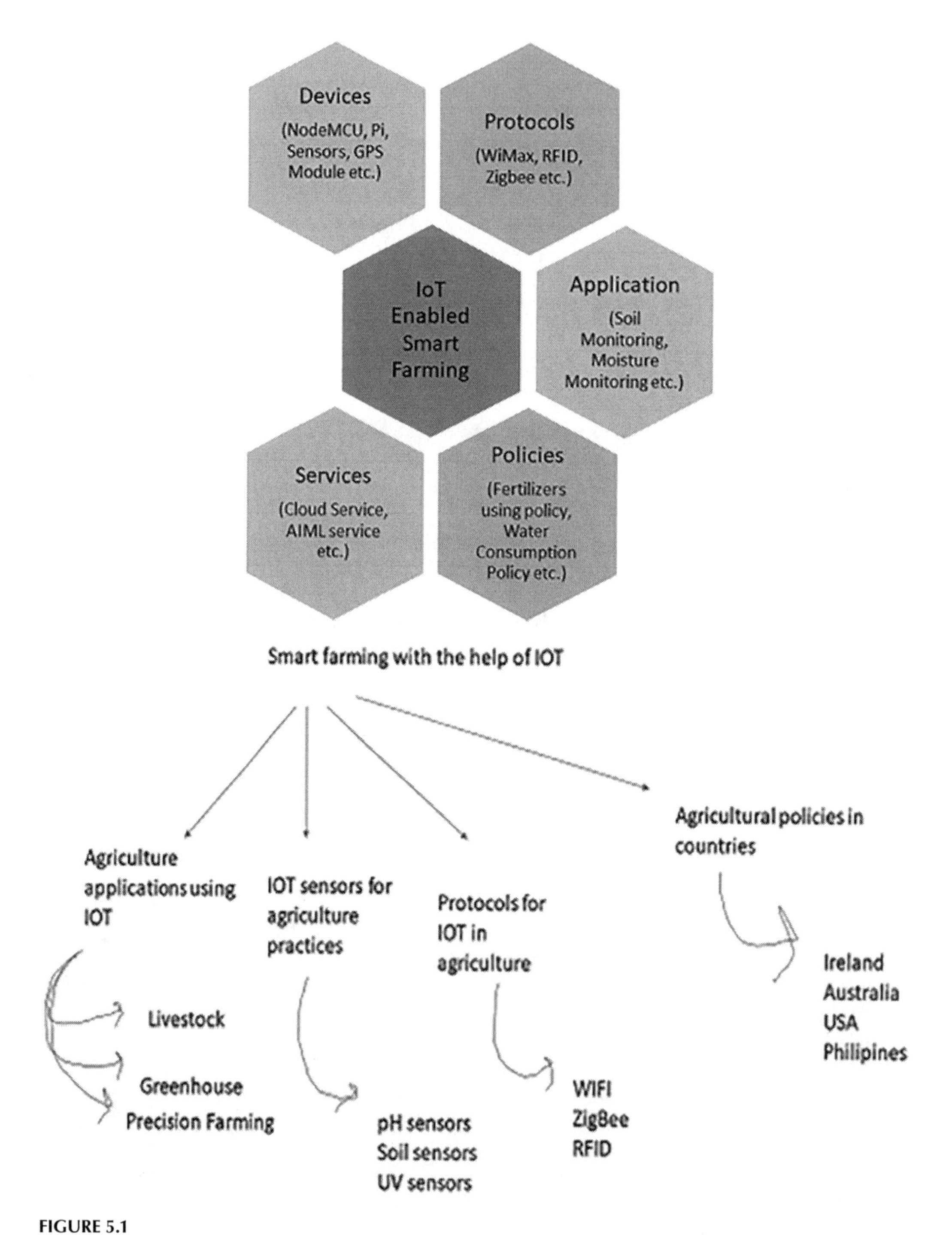

FIGURE 5.1
(a) Five components required for IoT enabled smart Farming (b) Different types of sensors and protocols used for smart agriculture.

advancements such as robots for assembling and segregating fields, as well as the internet of things and sensor systems (Fig. 5.1).

5.2 IoT-Driven Precision Agriculture

This section presents a solution that uses a smart irrigation system, which takes environmental information and determines where and when appropriate irrigation is needed. Automation makes the work of the farmer much simpler and easier. Controllers in the system will monitor soil moisture content, temperature, humidity, etc. at the physical field site to automatically schedule the irrigation process. The system is implemented by IoT, which comprises microcontroller devices like Raspberry Pi, NodeMCU, sensors, water pumps, etc.

Some advanced technologies made revolution in traditional agricultural activities, including the internet of things, AI, Meta Language, robotics, etc. These latest advances, especially IoT, not only give a way to deal with issues more promptly and check and control improvement factors, like water framework, water, and soil safeguarding and cut off the use of fertilizers on an estate just to indisputably the base necessities, simultaneously, also changes how we see agriculture. To meet the necessities of the growing population and to get the most yield from their properties, farmers are moving to new developments energized by the internet of things. New "sagacious developing" applications, including IoT advancements, will enable the agriculture business to diminish waste. IoT-engaged precision agriculture procedures give farmers valuable instruments to improve each undertaking. These advancement-driven practices revolve around extending crop yields and advantages while cutting down the levels of standard information sources (water, excrement, bug toxins, and herbicides) that are relied upon to foster harvests. For example, GPS devices on work vehicles enable farmers to plant crops more effectively and smooth out movement over and between their fields, saving time and fuel [3]. Sensors presented on developing components can moreover accumulate data related to the environment, soil, bug, or hydration conditions, then send that data to a fused smart farm stage to take apart and make judicious developing decisions. Fields can be leveled out by IoT-controlled lasers so that water can be applied even more beneficially and with less liquid waste running off into close by streams and streams. PC-based knowledge can in like manner be used for anticipating problems, which can be valuable for early action for bug control. Useful vermin the board prompts lower harvest and normal damage to the crop yield [4, 5]. A mix of indirectly distinguished data, useful picture portrayal gadgets, environment data, and other critical data centers can be used to identify the weed from the crop. This will confine the use of weedicides just to the spaces that require treatment. Far off satellites can screen crop prosperity and alert against bug attacks. Right when the chance shows up to give fields crop confirmation things, robots can be used rather than crop dusters to diminish expenses and wipe out the peril of controlled plane flying at high speeds so close to the ground. While flying, crop cleaning robots can take consistent photos and videos of fields so farmers can screen plant prosperity without sending out scouts.

The result of these precision agriculture procedures tends to an asylum for farmers to increase crop yielding to fulfill the need of the consumers.

5.3 Smart Farming to Get Optimized Irrigation

A precision water framework association set up by farmers in 2009, has encouraged a splendid development course of action that engages cultivators to suitably manage their yields' water framework. They start by perceiving a field's change using estimations, for instance, soil type, surface, topography, inclination, and yield (Fig. 5.2).

IoT-enabled agriculture has executed current mechanical responses for reliable data. It has moreover overcome any hindrance among crop production and quality and sum yield. With reliability from beginning to end, brilliant exercises, and improved business measure execution, produce gets arranged faster and shows up at stores in the quickest time possible. Aeris IoT gives precision agriculture organizations the advantage of changing separate things into related courses of action to achieve ideal resource use and usefulness in any field, wherever in the world [6].

5.4 Wireless Sensors in IoT

Wireless sensor networks that use improvised framework organization have become an area of intense assessment activity. This is due to the presence of inexpensive sensors for identification, monitoring, and concentrating progress in information processing using detectors, wireless connections, and frameworks. Moreover, smart agriculture should use modern developments, in particular IoT and wireless sensors that are widely used to boost farm production. The internet of things is today an advanced view that consolidates the internet and everything. We may argue that the internet of things is open progress. Sensor devices are a key component of IoT, notably in the field of farming, that may be used cleverly from a developing point of view to refer to portable recognizers, actuators, canny sensors, etc. These sensors can produce a huge amount of agricultural data, which enables us

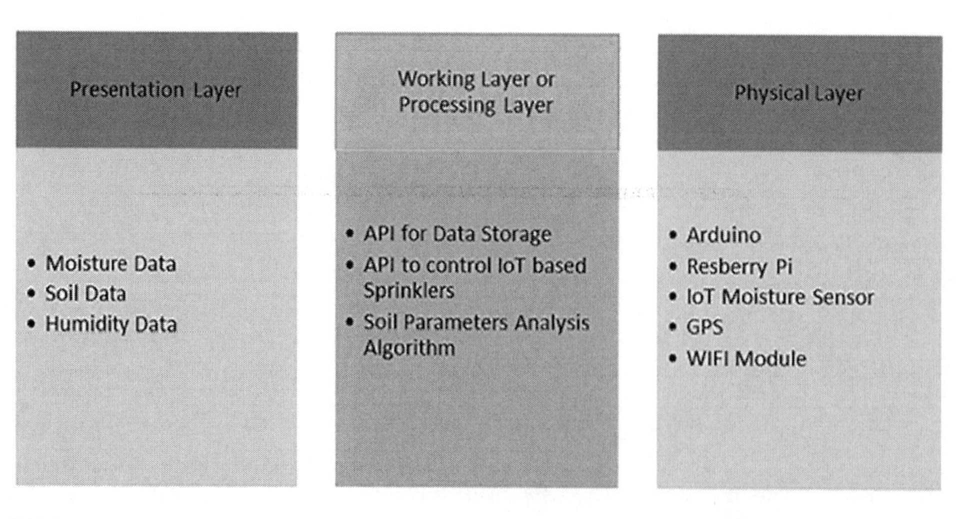

FIGURE 5.2
IoT-based irrigation architecture.

to obtain important and crucial information about evaluation, decisions, and intervention [7]. In view of the gigantic size of a sensor assignment, coordinating the message from a sensor center to a sink center is difficult. Data routing is therefore a crucial issue for wireless sensors. Typical guiding norms are floods and chirping. Calculations for AIs may be used to anticipate (e.g., when a machine falls), perceive (e.g. irregularities, water spills, etc.), and inspect in order to assemble, summarize, isolate, and collect sensor data. The easy test is to define a standard for correspondence with all heterogeneous rapid residence items. Cloud platforms allow sharp property objects to communicate data that is dynamically organized in the cloud, to restrict and concentrate critical information from sizable farm data, and to execute unprecedented hardware stages.

5.5 The Smart Farm

The wider population has certainly contributed to the promotion of agriculture. The key test for agriculture is to improve agricultural skills and quality without manualizing the efforts to meet food demand quickly. In agriculture too, natural change is a troubling model. In fact, to achieve farm advantages and outstanding output improved productivity the best test in quality development is the unconventionality of the environment.

In addition, today's agriculture, human manual intervention often has different obstacles, particularly in relation to continuous maintenance, in which the correct mediation may save a significant cost and an erroneous response can be excessive (Fig. 5.3). In particular, in regard regular circumstances, ongoing manual endeavors proved unreasonable and not practicable in great detail. IoT gives some kind of assistance in the interface of wireless sensors and other gadgets. As a result, the structures are resourced for intelligent items (e.g., temperature sensor, strain sensor). These days, agribusiness can exploit the growth of the network. They are combined with the techniques of recognizing that various intellectual farm sensor data, supported by Web applications, are permitted consistently and far away to help the agricultural region provide a secure, sound, sustainable livelihood. In addition, wireless sensors can be developed in open-source sensor-based applications that are less extreme and feasible for agricultural use. In the cloud environment and with the flexibility to adapt farms and apps to maintain their cycle via mobile and work stations, back-end data for intelligent inspection and careful maintenance development may be taken into account. Wireless sensors are important to intelligent development in order to enhance cultivation [8, 9]. At this moment, sensors can intervene in most of the provinces. Data obtained from the sensors, to modify development practices and increase farm safety and stability should be thought through for a fascinating and substantial investigation. Take an overview of a building with a crazy area of infection, a stupid hot zone with crop extraction devices. IoT intervenes to ensure that the yield extraction method remains acceptable. Next to a farm equipment center temperature and vibratory sensors and robots are set up. The following table demonstrates how IoT and sensors can guarantee maintenance of the system. In this manner a water system framework that controls the stream according to the prerequisite alongside computerization in the water system framework is planned and accomplished agreeably. With the utilization of minimal effort, sensors and straightforward hardware, which can be purchased even by a poor agriculturist, make this task easy. This work is most appropriate for places where water is scarce and must be utilized in constrained amounts. Information can be transmitted utilizing radio

FIGURE 5.3
Smart farming architecture devices.

recurrence signals in which every valve is furnished with a radio receiver and unraveling circuit. Here information (code to recognize every valve) is communicated and received by each valve and decoded, and only the valve having same code can be turned ON.

5.6 Implementation of Smart Irrigation System Using IoT-Based Smart Sensors

Agriculture is one of the most important sectors in the world. A proper irrigation process is essential for the cultivation of crops. Many farmers still perform the process through manual control, which requires a lot of hard work and results in wastage of water and power resources. Beside these problems, farmers face unpredictable changes in weather conditions and unavailability of enough water. Therefore, we require an intelligent automated system having the capability to precisely monitor and control the water and energy consumption. Nowadays automation plays a dominant role in human life. We implemented a smart automated irrigation system to overcome the shortcomings of traditional systems like pot irrigation or drip irrigation, which result in soil erosion and water wastage. An automated irrigation method distributes water to the crops by spraying like a natural rainfall. Installing a smart irrigation system saves time and ensures judicious usage of water. We designed a system using IoT technology devices like Raspberry Pi, NodeMCU, relay modules, and various sensors, etc. and ensured that the crops are watered appropriately when required [10]. The water flow will be monitored and predicted based on the data available and analysis. This will help farmers to manage water and power, indirectly reducing the cost and increasing the efficiency.

5.6.1 Module 1: Study of Requirements for the Work

Identifying the requirement is a vital a part of any research work. Understanding what an implementation work can deliver is vital to its success. Necessities gathering looks like wisdom, however amazingly, a neighborhood has given so much deficient attention on collected data items. Requirements included software as well as hardware. Hardware requirements basically included 2 GHz x 86 processor or above, 256 MB of system memory (RAM) or above, 100 MB of hard-drive space or above, a monitor to display output, and a keyboard/mouse for data input. Software requirements included a C-compiler (CC, GCC, EGCS), MS-word (for documentation), and Arduino IDE [11–13]. The basic architecture of IoT based smart irrigation system is shown in Fig. 5.4.

5.6.2 Module 2: Identification of Land and List of Equipment

After all the requirements were studied, then came the most important thing, which was where we were going to implement our work. For that we required land where we could perform a small-scale implementation of the smart agricultural scheme.

After the land was finalized, we moved onto the various devices/equipment that we required during the work implementation phase. Devices are listed as follows:

1. Raspberry Pi 3 Kit
2. Display screen for pi
3. Solenoid valve
4. Rain pipes for sprinklers work
5. Node MCU ESP – 8266 (Fig. 5.5)
6. Soil Moisture Sensor (Fig. 5.6)
7. Temperature and Humidity Sensor (DHT 11) (Fig. 5.7)
8. Relay Board (12V)
9. Protoboard/Breadboard
10. Relay for water flow
11. PVC Pipes
12. PVC Connections
13. Electricity wires and connections
14. Water electric pump

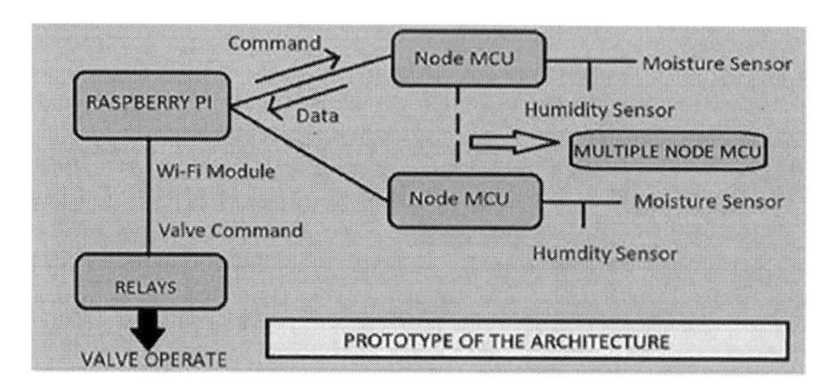

FIGURE 5.4
Architecture of the IoT-based smart irrigation system.

15. Pressure pump
16. Motor

5.6.3 Module 3: Building of the System and Connecting Equipment

After the equipment was finalized a small prototype of the architecture was prepared, showing how the system is going to be and how all connections will be made (Fig. 5.8).

FIGURE 5.5
Hardware component (Node MCU ESP – 8266).

FIGURE 5.6
DHT11 and soil moisture sensor connected to node MCU with water sample.

FIGURE 5.7
Raspberry Pi connected.

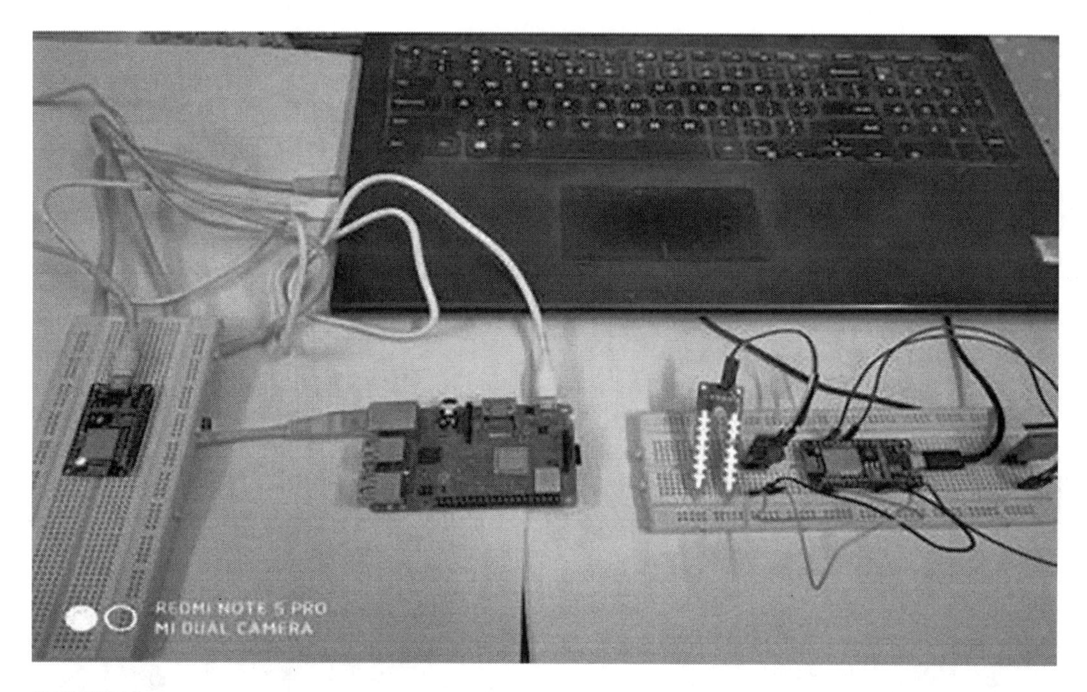

FIGURE 5.8
Raspberry Pi connected to NodeMCU.

5.6.4 Code for Arduino Integrated Development Environment (IDE)

Figure 5.9 displays the programmatic logic that will help us get the values from the multiple sensors so deployed. This piece of code includes prebuild header file "DHT.h," which basically contains various functions in order to record the values from the sensor and then pass it onto another function which will save the recorded values in a .txt file (Fig. 5.10).

This output displays the results of the Arduino IDE. The first thing to be displayed is the analog signal showing whether sprinklers need to be ON or OFF. Then humidity in the soil at that particular area, measured in percentage; temperature of the particular region, measured in degrees Celsius; and moisture content, also recorded in percentage.

Figure 5.11 displays the programmatic logic that will help us store the recorded values from the Raspberry Pi in a .txt file.

5.6.5 Saving Data in a Text File

Figure 5.12 displays the result of the programmatic logic defined in Figure 5.11.

5.6.6 Module 4: Deployment of System on Identified Land and Results

Once the testing is done; the work will be deployed on the identified piece of land (Figs. 5.13–5.15).

FIGURE 5.9
Arduino IDE coding console.

FIGURE 5.10
COM3 serial port monitor.

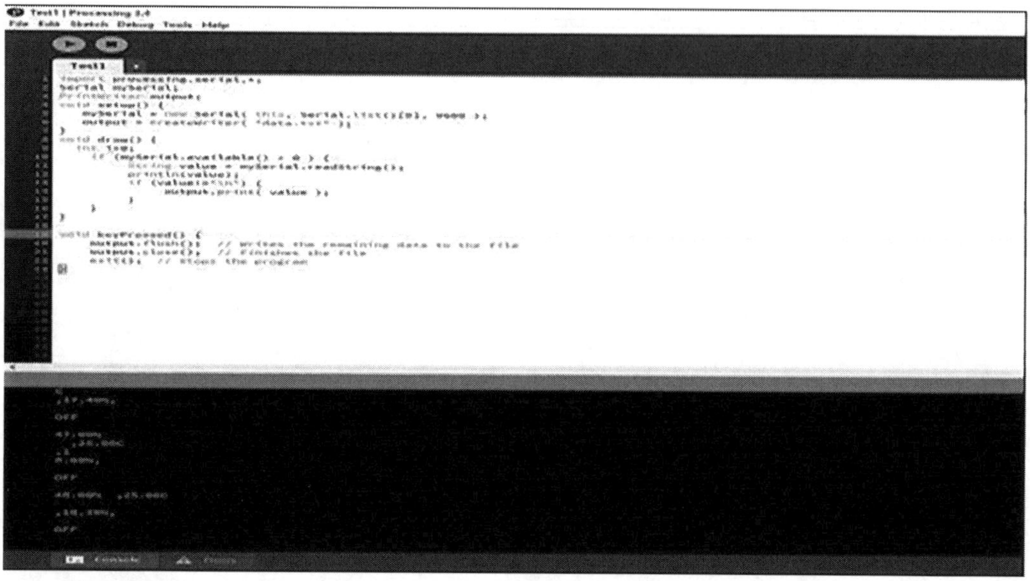

FIGURE 5.11
Code for storing data in text file.

```
data - Notepad
File  Edit  Format  View  Help
ON45.00%   ,24.00C
,25.32%,
ON
```

FIGURE 5.12
Data in text file.

5.7 Conclusion

The execution of IoT structures in rural locales may present a couple of challenges due to the vegetation. Particularly, agriculture applications relating to soil quality may encounter negative impacts of the thickness of the foliage or a couple of qualities of the plants, such as height or width. In this paper, we have proposed an earth-checking system that incorporates a soil dampness multi-sensor group, an earth temperature sensor, and a pH sensor, and we have played out an assessment on fields with different arrangements with varied

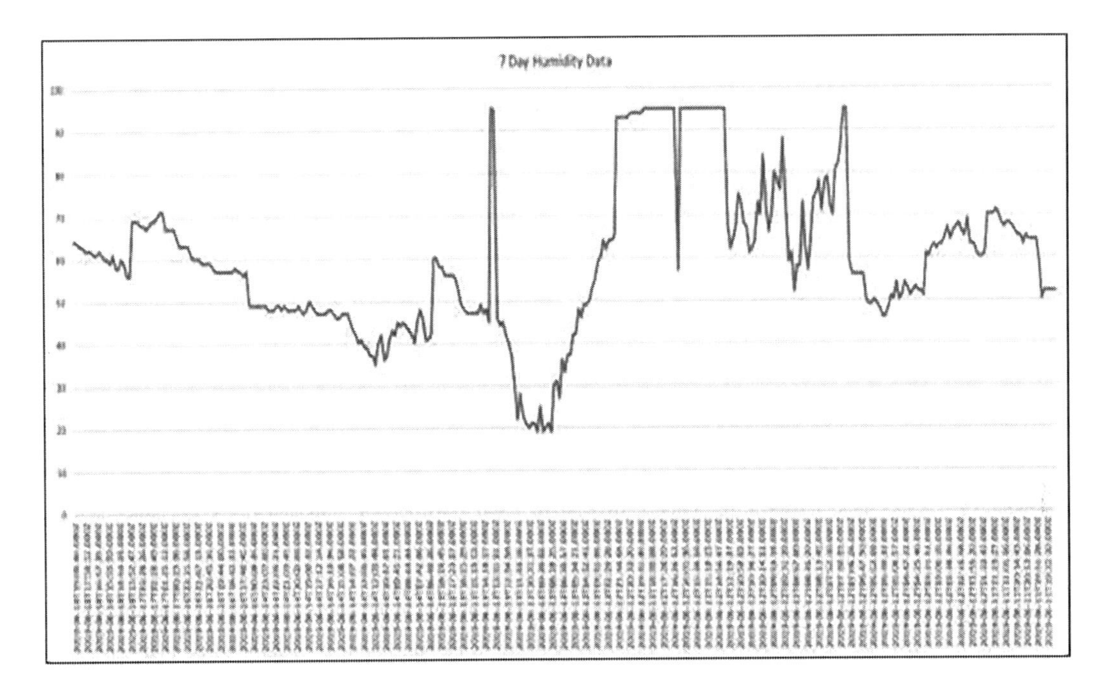

FIGURE 5.13
Data from humidity sensor.

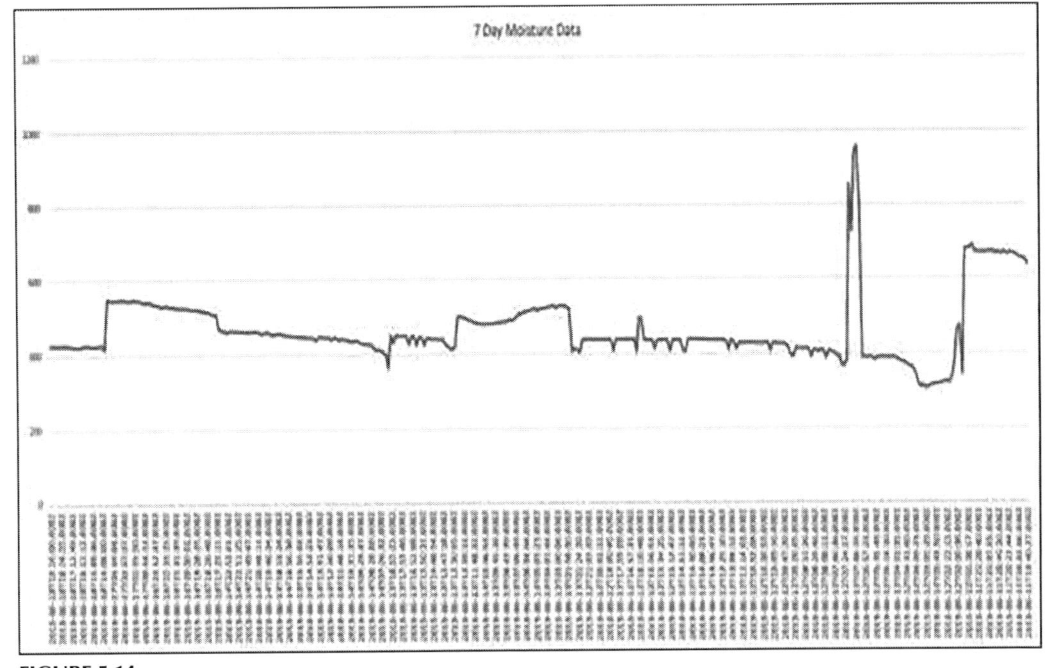

FIGURE 5.14
Data from moisture sensor.

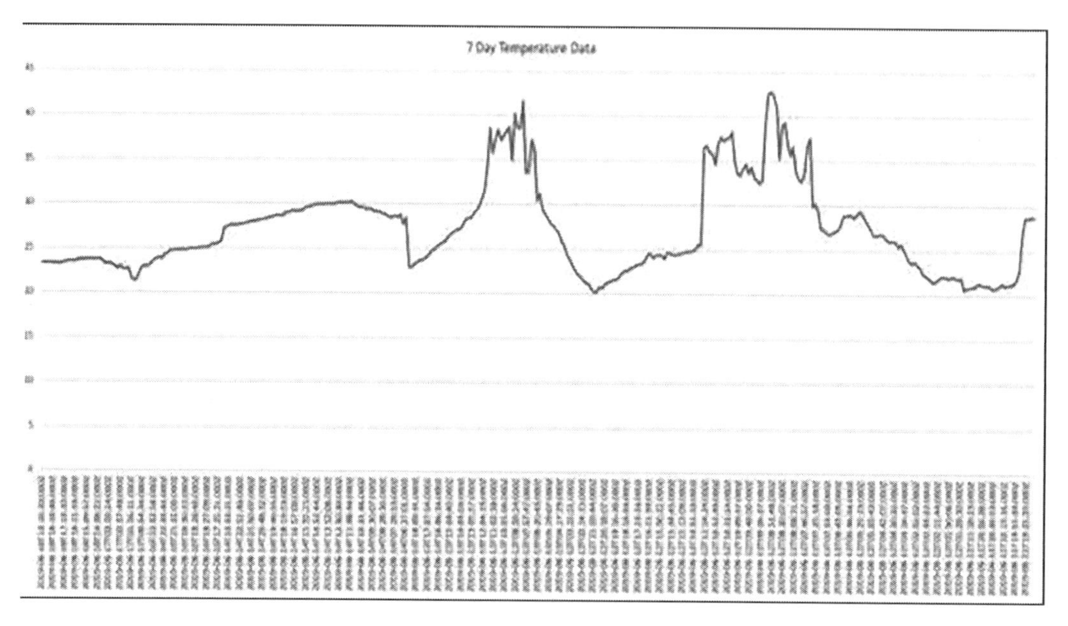

FIGURE 5.15
Data from temperature sensor.

sorts of vegetation. Also, on-ground, close-ground or more ground-center plans were attempted as well. Results have shown that vegetation varies greatly in districts with high foliage thickness. As we know, the expected population of India is 1500 million by 2050, and agriculture will be the primary source of livelihood in rural areas, so the focus should be on increasing the productivity of crops without wasting another precious commodity: water. Though our country claims to be developed in terms of science and technology, an erratic power supply has become almost routine. India has a huge untapped solar off-grid opportunity and also receives heavy rainfall. So why not utilize what we have to increase the crop growth of our country and take it towards prosperity?

What about the irrigation system? Is it smart enough? In our research we designed architecture for automatically controlling the irrigation system. We use multiple sensor nodes (soil moisture and DHT11) and a control node [14]. The sensor nodes are deployed in the field so that they can collect respective values, and the sensed data is then sent to the controller node. On receiving the data, the controller node checks the sensed value with a set threshold value. The threshold value depends on the type of crop being cultivated. When the soil moisture value or temperature/humidity value is not up to the required level, the controller node sends instructions to switch the motor ON. The automated system is capable of automatically irrigating the field based on collected and sensed environmental data surrounding the crop. Moreover, the system can be monitored and controlled remotely due to the use of a WiFi module present in the NodeMCU. Moreover, the power consumed by wireless network devices is also less, and the system performs its function for a longer duration of time. This is a great advantage to the farmers who want to monitor the field remotely. This saves a lot of time and energy. Optimization of water usage is also achieved.

The most outrageous speculative consideration was gotten for each plan. On-ground associations had the least incorporation even with vegetation where most of the foliage is higher. In any case, the elements of the common environment and the association that impact the sign like center stature, crop type, foliage thickness, or the kind of water framework ought to be seen as when arranging a WSN course of action for PA structures as it has been analyzed.

References

1. Farooq, M. S., Riaz, S., Abid, A., Umer, T., Zikria, Y. B., "Role of IoT technology in agriculture: a systematic literature review", Electronics, 2020, 9(2), pp. 319. https://doi.org/10.3390/electronics9020319
2. Marcu I. et al. (2019) Overview of IoT Basic Platforms for Precision Agriculture. In: Poulkov V. (eds) Future Access Enablers for Ubiquitous and Intelligent Infrastructures. FABULOUS 2019. Lecture Notes of the Institute for Computer Sciences, Social Informatics and Telecommunications Engineering, vol 283. Springer, Cham. https://doi.org/10.1007/978-3-030-23976-3_13
3. https://www.krishisewa.com/miscellaneous-articles/1018-use-of-iot-based-irrigation-scheduling-for-smart-farming.html
4. Dasig D.D. (2020) Implementing IoT and Wireless Sensor Networks for Precision Agriculture. In: Pattnaik P., Kumar R., Pal S. (eds) Internet of Things and Analytics for Agriculture, Volume 2. Studies in Big Data, vol 67. Springer, Singapore. https://doi.org/10.1007/978-981-15-0663-5_2
5. Alli, A. A., Alam, M. M. "The fog cloud of things: a survey on concepts, architecture, standards, tools, and applications", Internet Things, 2020, 9, 100177.
6. https://labs.sogeti.com/iot-vs-edge-vs-fog-computing/
7. Ahlawat, P. et al., "Sensors based smart healthcare framework using internet of things (IoT)", International Journal of Scientific and Technology Research, 2020, 9(2), pp. 1228–1234.
8. Ahmed, E. et al., "I-Doctor: An IoT Based Self Patient's Health Monitoring System", 2019 International Conference on Innovative Sustainable Computational Technologies, CISCT 2019 (2019).
9. Agarwal, N. et al., "Real time activity logger: a user activity detection system", International Journal of Engineering and Advanced Technology, 2019, 9(1), pp. 1991–1994.
10. Sharma, H.K. et al., "Air Quality Prediction using Artificial Neural Networks", 2019 International Conference on Automation, Computational and Technology Management, ICACTM 2019, 2019, pp. 568–572, 8776774.
11. Sharma, A. et al., "Automated parking system-cloud and IoT based technique", International Journal of Engineering and Advanced Technology, 2019, 8(4C), pp. 116–123.
12. Kumar Sharma, H., Kshitiz, K., Shailendra, "NLP and machine learning techniques for detecting insulting comments on social networking platforms", Proceedings on 2018 International Conference on Advances in Computing and Communication Engineering, ICACCE 2018, 2018, pp. 265–272, 8441728.
13. https://medium.com/yeello-digital-marketing-platform/what-is-fog-computing-why-fog-computing-trending-now-7a6bdfd73ef
14. Anand, A. et al., "Traffic management in MPLS network using GNS simulator using class for different services", International Conference on Computing for Sustainable Global Development, INDIACom 2015, pp. 1066–1068.

6

A Study of Blockchain-Based Energy-Aware Intelligent Routing Protocols for Wireless Sensor Networks

Arup Sarkar and Geetansh Atrey

CONTENTS

6.1 Introduction

A wireless sensor network (WSN) (Fig. 6.1) is a network of devices or a group of distributed sensors that communicates information gathered from a monitored area through wireless links (infrastructureless), thus organizing the entire data collected at a central location. It is generally used to monitor and collect information about physical conditions of the environment where the sensors are present. Some common parameters measured using WSNs are temperature, noise levels, wind, humidity, etc. They have numerous other applications in the fields of agriculture, battlefields (for threat detection), and medical appliances, and

DOI: 10.1201/9781003102397-6

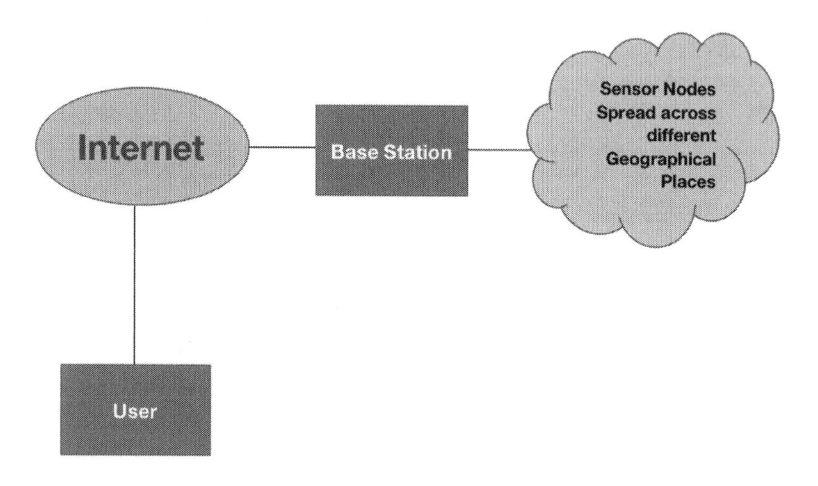

FIGURE 6.1
WSN.

also in the internet of things (IoT). This network of sensors is deployed in an ad hoc manner. WSNs consist of four components: sensors, radio nodes, Wireless Local Area Network (WLAN) access point, and evaluation software.

i. Sensors: WSN's sensors are used to capture environmental variables and to collect data from the environment, such as humidity, temperature, noise level, etc. The sensor signals are converted into electrical signals for further transmission.
ii. Radio nodes: The nodes receive the data collected by the sensor and forward them to the WLAN access point. Radio nodes contain a variety of components such as external memory, micro-controllers, transceivers, and power sources.
iii. WLAN access point: The information that is sent by the radio nodes in a wireless manner is received by WLAN access point, which generally is done by using the internet (network of networks or inter-network).
iv. Evaluation software: Evaluation software can be used for many useful purposes, such as data processing, analysis, storage, and mining of the data obtained by WLAN access points that can process data for submission to users for further use and analysis.

With a base station, sensor nodes communicate the data collected from the environment for forwarding and for storage purposes. These sensors are scattered, so WSNs have some challenges and constraints in limited bandwidth, node costs, deployment, design, and time synchronization. But if we look into it, the biggest challenge faced by WSNs is the energy consumption, or power consumption. Sensors require energy or some form of power backup constantly in order to perform their operations. These nodes generally have batteries with limited power capacity. But this power is consumed in a lot of processes, such as sensing, data collection, processing, and transmitting the data recorded from the environment to the sink nodes. According to Figure 6.2, most of the consumption of energy (around 60%) takes place for communication of data between the sensor nodes and the sink nodes (formally known as a communication subsystem). Around 25% energy consumption is needed for the computing subsystem, which includes the micro-controllers and data storage. The remaining 15% of energy is consumed by the sensing subsystem, which basically consists of the sensing mechanism, including sensors, and converting sensor signals to electrical signals.

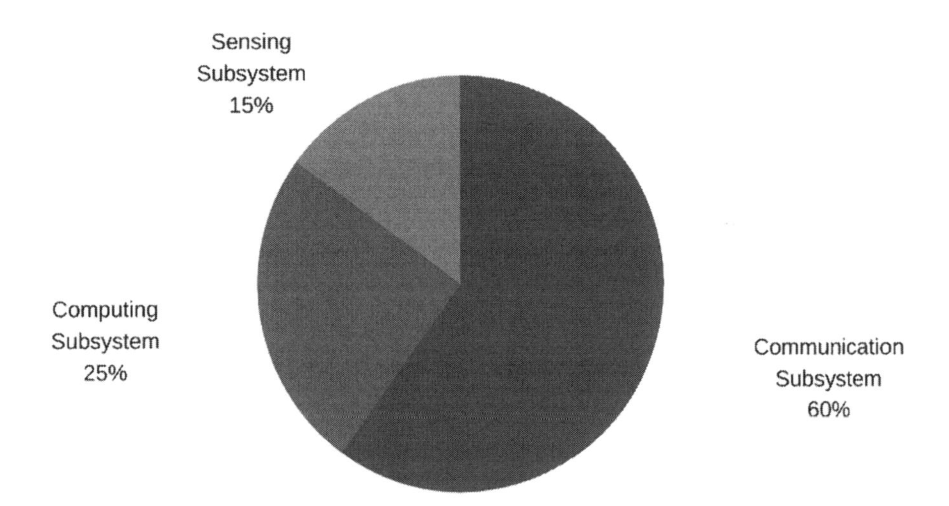

FIGURE 6.2
Distribution of power consumption for a wireless sensor node.

Analyzing the level of power consumption by different processes shows that most of the energy is consumed in the transmission process. For the same energy cost of sending a single bit 100m by radio, around 3000 instructions can be executed. At the same time, remote sensors also have to be active all the time for listing sink node or base station's queries. Thus while waiting for the same, nodes are not performing their task of monitoring the environment and thus are sitting idle. Thus this idle period also leads to some amount of wastage of energy. As shown below in Figure 6.3, most of the energy is consumed during transmit mode (Tx), receive mode (Rx), and idle operation.

The nature of the sensor nodes and the dangerous environmental conditions make it very difficult to replace the batteries. But there are many types of sensors that require a longer life. Thus energy consumption is a big issue in WSNs. The sensor nodes have irreplaceable and non-rechargeable batteries. Power efficiency is the key requirements to maximize sensor node lifetime. There needs to be an optimal balance between power consumption and energy storage. One possible solution for this is to develop batteries with high power capacities, but it is unlikely to help in this case because of slow progress in this field. Using energy-efficient routing algorithms and conducting T_x / R_x operations in specific situations can help address this problem.

Network routing can be defined as the process of selecting an optimal path across one or more networks in order to deliver a packet. Routing as a concept is a very necessary procedure in a lot of areas, right from telephone networks to public transportation. Selecting the optimal path is the main task achieved during routing, which in return leads to the entire process of transport of data being efficient. In packet-switching networks, for example, routing selects the optimal path of the internet protocol (IP) packets for travel from their source to their destination.

In the context of WSNs, routing becomes even more important. In WSNs, a routing technique is required for sending the information collected from the environment by the

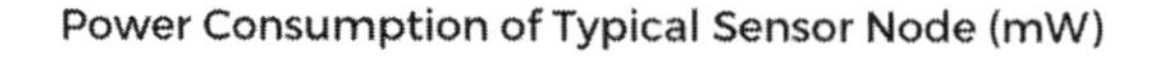

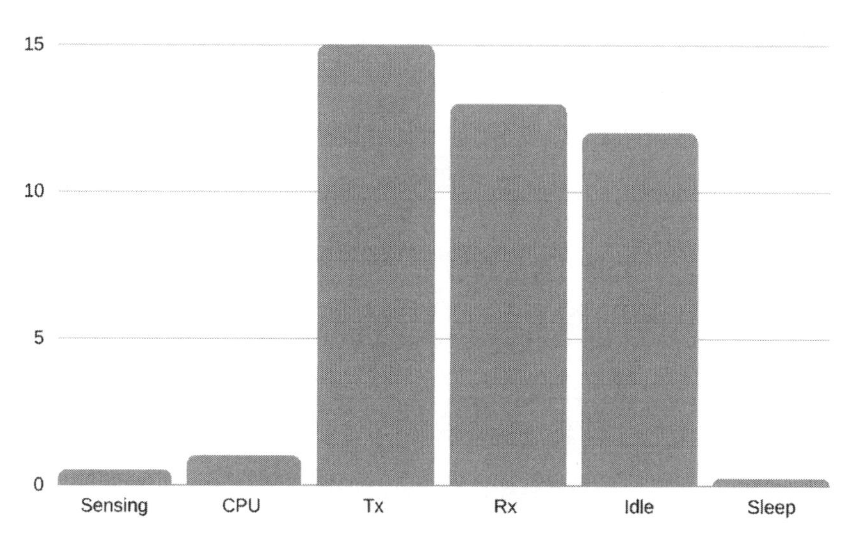

FIGURE 6.3
Consumption of power of a sensor node (mW).

sensor nodes to the base stations in order to establish communication. But at the same time, the process meets several challenges while selecting the optimal route, which is dependent upon some key factors, such as channel characteristics, the type of network, and the performance metrics. The data collected from the environment by the sensor nodes in a wireless sensor network is typically sent to the base station, which connects the sensor nodes with the entire network of sensor nodes (maybe network of networks – the internet), where the data is collected, stored, processed, and analyzed and some relevant actions are taken according to the calculated environment variables. In small networks of sensors, typically consisting of sensor nodes and base stations being very near to each other, communication can take place directly between them. This type of communication is known as single-hop communication. The single-hop communication is also called direct communication. But practically in most WSNs, the coverage area of the network is very large, consisting of thousands of sensor nodes with the base station being far from them. This requires multi-hop communication because the sensor nodes cannot communicate directly with the base station. Multi-hop communication is also known as indirect communication. Since multi-hop communication consists of thousands of sensor nodes, routing is effective. Thus in the case of multi-hop communication the sensor nodes not only perceive the environment but also act as a path for other sensor nodes towards the base station during routing. Of all the protocol layers, it is the network layer that is responsible for routing.

A reliable routing scheme is crucial for the high efficiency of the WSN and to ensure high security in routing. There are numerous plans to improve the reliability of routing nodes using trust management, cryptographic systems, or centralized routing decisions. However, most routing strategies are difficult to attain in real situations because the unreliable behaviors of the routing nodes are difficult to detect dynamically. In the meantime, there is still no effectual way to prevent the attack of malevolent nodes. With these issues in mind, in this chapter we will look at trusted routing schemes for blockchains and reinforcement learning to better routing security and efficiency for WSNs. Blockchain routing schemes will be discussed in order to obtain information about the routing of routing nodes in the blockchain, which makes it impossible to identify routing information identifiably and incompletely. The learning model is used to help routing nodes dynamically select more reliable and efficient routing links.

6.2 Types of Routing Protocols

Different categories of WSN routing protocols are available on the following basis: (1) functioning mode, (2) participation style, and (3) network structure (Fig. 6.4).

1. Functioning mode: The operation of the wireless sensor network specifies its application. Thus, routing protocols can be classified according to the operation used to satisfy the WSN function:
   a. Proactive protocols: Data is transferred to base stations by default route. These protocols are also called table-driven protocols (for example, Optimized Link State Routing [OLSR]).
   b. Reactive protocols: In reactive protocol, routing is established according to demand. The route is set up dynamically. Network-based routes are available when needed (for example, TIN, AODV, DSR).
   c. Hybrid protocols: In hybrid protocols, all routes are initially identified and then corrected at the time of data transmission. These protocols combine and retain the concepts of both reactive and functional (for example, Apten).

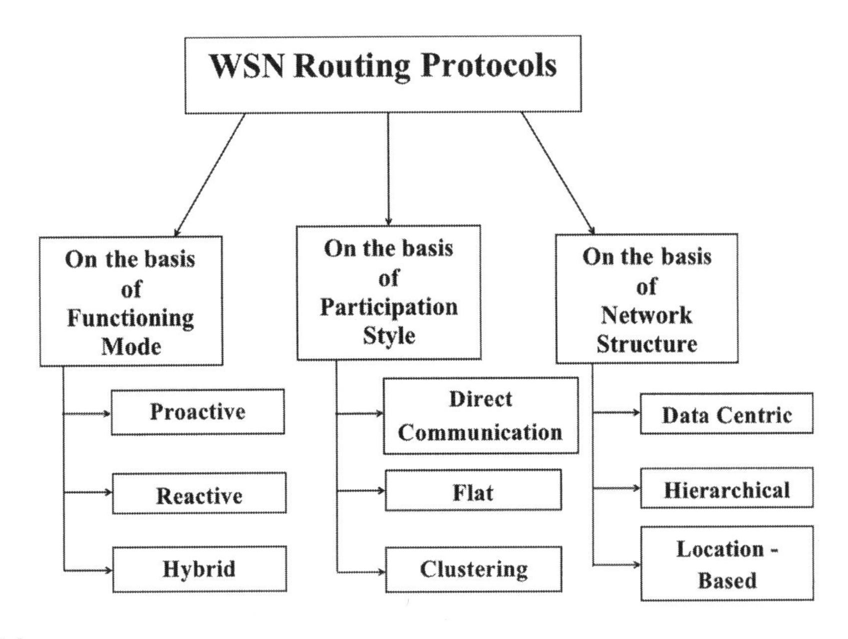

FIGURE 6.4
Different categories of energy-efficient WSN routing protocols.

2. Participation style: If we look at WSNs, some networks have identical nodes, while others have different nodes. Thus, both types of nodes participate in each network individually as different networks (e.g. remaining power of nodes, cluster heads, etc.). Based on this concept, we can classify protocols:
   a. Direct communication protocols: In direct protocols, the data collected by the sensor node are transmitted directly to the base station, since the distance between the node and the base stations is short (for example, the spin protocol).
   b. Flat protocols: The sensor nodes in this protocol search for the valid path for packets and then transmit it to the base station (for example, rumor routing protocol).
   c. Clustering protocols: In clustering protocol, the whole area of the network is divided into clusters, and each cluster is assigned a cluster head. All data collected by the sensor node are transmitted to the corresponding cluster heads, and then the cluster head forwards them to the base station (for example, the TEEN network-based routing protocol).
3. Network structure: Network-based routing protocols depend on the structure of the network, its layout, or the strategies of how the network is organized. This national protocol falls into three categories:
   a. Data centric protocols: Data-centric protocols are based on queries and they rely on the naming of desired data. Base stations send queries to specific geographic areas of the network to receive information and wait for sensor node responses. Sensor nodes in a specific region collect specific data based on the query (for example, the SPIN protocol).
   b. Hierarchical protocols: In hierarchical protocols, the sensor nodes are separated by the amount of energy they contain. Nodes with low power are used to receive information from the environment, and nodes with high power are used for further processing and transfer to the base station. This is why such protocols are considered for efficient routing (for example, TEEN, APTEEN).
   c. Location-based protocols: In location-based protocols, flood is used to find the best path with the location of already known nodes. GPS is used to obtain information about the location of these nodes (for example, GEAR).

6.3 Energy-Aware Intelligent Routing Protocols for Wireless Sensor Networks

The cognition of network construction and routing protocol is very important in respect to WSNs because both the structure and routing protocol should be relevant for the requirement of the usage. Some of the important and commonly used energy-aware intelligent routing protocols are as follows:

I. Low-energy adaptive clustering hierarchy (LEACH) [1]: LEACH is an adaptive, self-organizing, clustering routing protocol that organizes the cluster in such a way that energy is evenly distributed and loads are evenly distributed

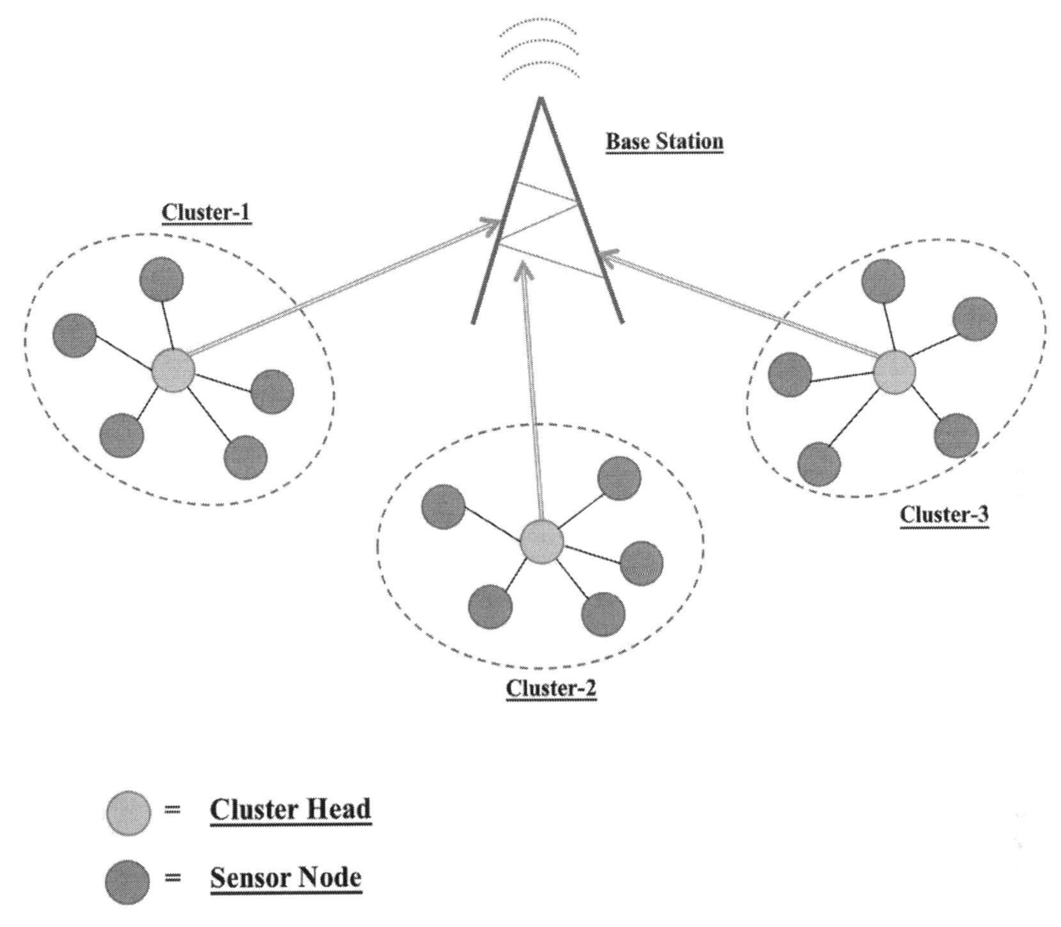

FIGURE 6.5
LEACH protocol and its work using clusters.

across all sensor nodes in the network. The idea of clustering comes from the LEACH protocol. As shown in Figure 6.5, many clusters originate from multiple sensor nodes where one node is defined as a cluster head or local base station and thus acts as a routing node for all other nodes in the cluster. Cluster heads are selected before full communication begins. If a situation occurs where no cluster head is selected or there is a problem with the cluster head, communication fails and the battery is more likely to drain faster than other nodes in the cluster. The head of the cluster is responsible for routing for the entire cluster.

Thus, the selection of the high-power cluster-head position involves LEACH randomization so that different sensor nodes can achieve the position of the cluster head without draining the battery of a single sensor. Some probability criteria are defined by the protocol according to which the sensor nodes select themselves as cluster heads. Nodes that have already become cluster heads may not exist again for the X round, where X is the desired percentage of cluster heads. Thus, each node has a 1 / X probability of becoming a cluster head again. At the end of each round, each node that is not a cluster head selects the nearest cluster head and joins that cluster. It is then the responsibility of the head of the cluster to create a schedule for sending data to each sensor node's own cluster.

II. Power-efficient gathering in sensor information system (PEGASIS) protocol [2]: PEGASIS is the most remote, classified protocol based on the chain of choice. In this protocol, nodes are organized as a transformation chain and the data setup of the chain can be centralized depending on the application. PEGASIS is based on the idea that all nodes are positioned by global network awareness.

The protocol presentation of PEGASIS is shown below. Chain construction starts from the end of the sync node and is chosen as the next node of the neighboring chain in its vicinity, etc. The node acts as the head of the node before sinking, and the sink node must be the last node. Due to the periodic topology (or dynamic network behavior), Pegasus is not such an important routing protocol; operations like data processing and data aggregation are performed by the leader node.

The PEGASIS algorithm has just been submitted to the LEACH protocol. In PEGASIS, the basic principle is to create a chain between all the sensor nodes. So that the power node can be collected and transferred from the nearest neighbor. The aggregated data is transmitted from node to node and aggregated, and then the aggregated data is transmitted by the head of the cluster to the BS (base station).

III. Threshold-sensitive energy efficient sensor network protocol (TEEN) [3]: Although LEACH was a practical protocol, TEEN was targeted at responsive networks and was the first protocol created for truly responsive networks (Fig. 6.6). In TEEN, a sensor node senses the environment continuously but changes the value of the sensor only when the radio is turned on. There is no periodic transmission of data. In this scheme, during the change of each cluster, in addition to the required features, the cluster head also transmits to its member nodes, hard threshold and soft threshold.

i. Hard threshold (HT): This is a threshold value for sensitive properties. This is the absolute value of the attributes that the transmitter of the node that understands this value should be turned on and reported under the heading of its cluster.

ii. Soft threshold (ST): This is a small change in the quality of the senses. Whenever there is a small change, it triggers its transmitter and the node to transmit.

The node turns on its transmitter and transmits sensing data for the first time when any parameter from the attribute set exceeds its rigid threshold value. Sensory values (SVs) are stored in an internal variable between nodes. During the current cluster, the nodes will send data only if the following two conditions of the organization are true:

1. The sensed attribute's current value is higher than the hard threshold.
2. The current value of the sensory property varies with respect to the SV by a greater or equal amount than the soft margin. Whenever a node transmits data, the current value of the SV sensitive multiplier is set equal to SV.

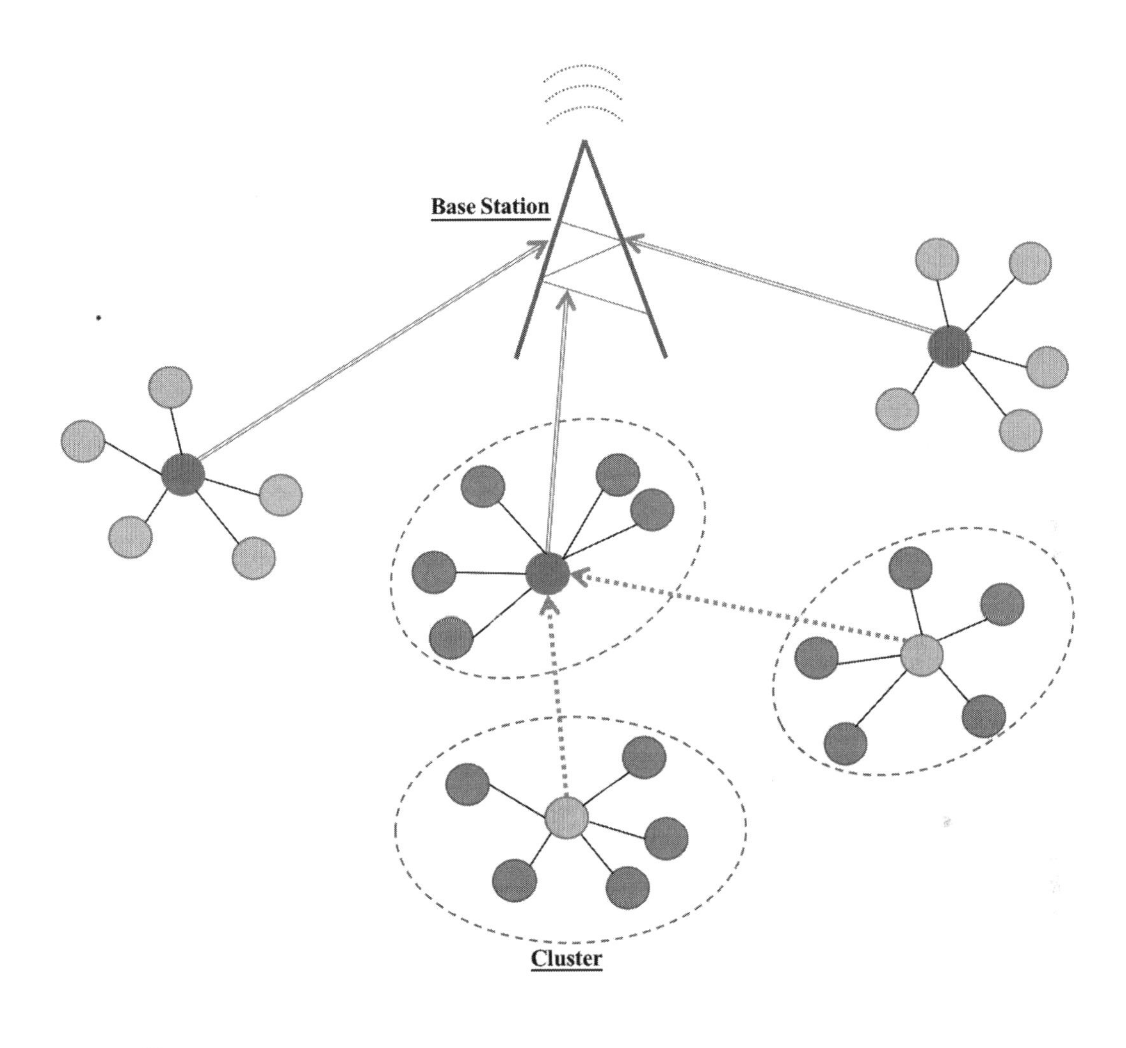

FIGURE 6.6
TEEN protocol and its work using clusters.

6.4 Performance Evaluation and Comparison of Various Energy-Aware Intelligent Routing Protocols for Wireless Sensor Networks

The following Table 6.1 compares the three energy-aware intelligent routing protocols for WSNs mentioned previously – LEACH, PEGASIS, and TEEN – with other well-known protocols.

TABLE 6.1

Tabular Comparison of Energy-Aware Intelligent Routing Protocols

Routing Protocols	Mobility	Power Management	Network Lifetime	Scalability	Resource Awareness	Classification	Data Aggregation	Query Based
LEACH	Fixed BS	Maximum	Very - Good	Good	Yes	Clustering	No	No
PEGASIS	Fixed BS	Maximum	Very - Good	Good	Yes	Reactive/ Clustering	Yes	No
TEEN	Fixed BS	Maximum	Very - Good	Good	Yes	Reactive/ Clustering	Yes	No
SPIN	Supported	Limited	Good	Limited	Yes	Proactive/ Flat	Yes	Yes
APTEEN	Fixed BS	Maximum	Very - Good	Good	Yes	Hybrid	Yes	No
RR	Very limited	Not support	Very - good	Good	Yes	Hybrid/Flat	Yes	Yes
DD	Limited	Limited	Good	Limited	Yes	Proactive/ Flat	Yes	Yes

TEEN, LEACH, PEGASIS, and APTEEN have similar characteristics and are comparable to some degree in their architecture. They have infrastructure that is fixed. LEACH, TEEN, and APTEEN are routing protocols based on a cluster, while PEGASIS is a protocol based on a chain. With regard to energy usage and network durability, APTEEN's efficiency lies between TEEN and LEACH. TEEN transfer time-critical data only, while APTEEN transmits data daily. In this respect, since APTEEN transmits data based on a threshold value, APTEEN is also better than LEACH, while LEACH transfers data continuously. Again, PEGASIS avoids the creation of LEACH overhead clustering but requires dynamic topology improvement because there is no monitoring of sensor energy. For distant nodes in the chain, PEGASIS causes unnecessary delay. In PEGASIS, the single leader can become a bottleneck. Compared to the LEACH protocol, PEGASIS increases network life two-fold.

The base station sends inquiry to the sensor nodes in the information directed by the flood strategies, but the sensor nodes publicize the availability of data in the SPIN so that curious nodes can ask for data. Each node can pass on with its neighbors in a guided diffusion, so it does not require full network data, but SPIN retains a global network topology. In the case of floods, the spin removes unnecessary data by half. Since SPIN cannot assurance data delivery, it is not appropriate for applications that necessitate reliable information delivery.

Metadata is used by SPIN, guided diffusion, and gossip routing, while the other protocols do not use it. Because they are flat routing protocols, routes originate in areas that include data for sending, but for others they form clusters across the entire network because they are classified routing methods. Energy dissipation is standardized and cannot be controlled in the case of classified routing, but in the case of flat routing the reduction of energy dissolution depends on the pattern of traffic. The cluster top is the aggregate data of the previous case, but the nodes of the multi-hop path combine the data from the neighbors in the latter case.

6.5 Blockchain and Its Need in Wireless Sensor Networks Routing Protocols

Blockchain can be defined as "a decentralized computing and information sharing platform that enables multiple authoritative domains, which do not depend on each other, to collaborate, coordinate and collaborate in logical decision-making processes." This definition is divided into several basic terms that define the characteristics of blockchain technology:

1. Decentralized: It means that there are multiple points of coordination in the network. That means that the network has no one central server that guides all the other nodes in the network. Instead, there are multiple small-scale servers connected to each other and thus no single point of failure.
2. Computation and information sharing: Blockchain provides a highly secure platform to share information among all the nodes present in the network. This can be any type of information from bank transactions to some public data. This is strongly backed by cryptography.
3. Do not trust: This basically refers to the fact that the nodes in the blockchain network don't know each other personally and hence can't trust each other while taking any form of decision.
4. Cooperate, coordinate, and collaborate: This refers to the consensus process in blockchain. These nodes don't trust each other but still are able to cooperate, coordinate and collaborate together to find a common truth.
5. Rational decision-making process: The decision that is taken in the blockchain network is the final used product.

In simpler words, blockchain is a **decentralized**, distributed, public ledger [4]. Public ledger is a database of historical information which is made available to everyone. This historical information is used for future computations. Taking an example, this historical information might be a record of transactions for a bank. Now these might be used in the future to validate transactions.

At its initial level, a blockchain is literally a chain of blocks. The term blockchain consists of two phrases, viz. "block" and "chain," where "block" refers to digital info that is stored in a public database or "chain." Blockchain has no central authority; it is the idea of a democratic structure. The data contained in it are available to everyone and anyone to view, because blockchain is transparent and sustainable. Therefore, everything that is made up in the blockchain is crystal clear by nature, and everyone participating is responsible for their work.

Blocks accumulate data that sets them apart from other blocks. A block stores a specific code called a "hash" that helps to differentiate it from any other block, as people have different names to differentiate them from each other. Hash is cryptographic code that creates special algorithms. In fact, a single block of the bitcoin blockchain can store about 1MB of data. This means that a block can have several thousand transactions under one roof, depending on the size of the transaction.

Security and confidence issues are responsible for blockchain technology in many cases. New blocks are accumulated in linear and chronological order. This is because they are always connected to the blockchain. Looking at the blockchain of bitcoin, there is a place in the chain of each block called "height." As of August 2020, the block height was 646,132.22.2. Once the block is added to the chain, it is very difficult to change the content of the block. This is because each block has its own hash, as well as the block hash. Math functions create hash codes that convert digital data into a string of numbers and letters. If that information is edited in any way, the hash code will also change. As shown in the Figure. 6.7, due to the ocean effect on the blockchain, any small change in the content of the blockchain reflects a massive change in subsequent blocks

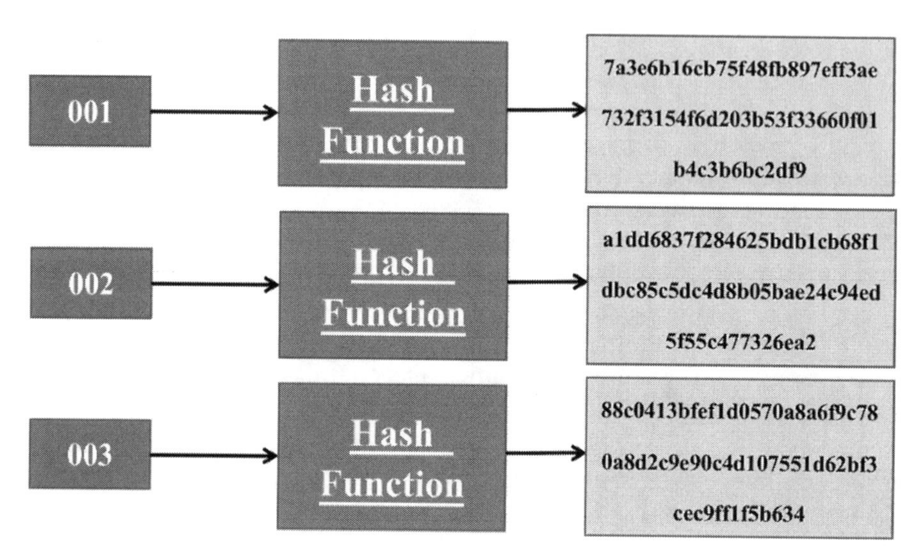

FIGURE 6.7
Avalanche effect in the hash values.

That is why protection is important. Suppose a hacker wants to perform a transaction that has to be done for a retail purchase, such as paying him twice for an order. The block hash will be adjusted as the transaction dollar sum is edited. The next block in the chain will still have the old hash, and the hacker must replace that block to cover their tracks. But doing so will remove the hash of that block, remove and next, and so on.

Then, to replace a block, the hacker must replace each block in the blockchain. To recalculate all these hashes requires a very high and unexpected amount of calculated power. In other words, once a block is added to a blockchain, it becomes very difficult to edit and impossible to delete. Blockchain networks have introduced checks for computers that want to enter the chain and add blocks to solve trust issues. This assessment, known as the "consent model," enables users to "prove" themselves before joining the blockchain network. A common example used by bitcoin is called "work proof."

Blockchain technology ensures security and avoids the need for trusted third parties for protection. However, implementing blockchain on a resource-limited wireless sensor network is a challenging task because blockchain demands a lot of bandwidth due to the power, calculation, and memory nature of hunger and control overhead. To ensure the routing security and efficiency of the WSN, a reliable routing scheme is very important. There are many studies to improve the reliability of routing nodes, use of cryptographic systems, confidence management, unified routing preferences, etc. However, in reality, most routing schemes are difficult to implement because the unreliable functions of routing nodes are difficult to detect dynamically. Meanwhile, there is still no successful way to avoid malicious node attacks. Current routing schemes seem to find it difficult to detect such malicious nodes, as it is difficult to accurately interpret the real-time change of routing information between two routing nodes. If a malicious node receives a data packet from a neighboring node, the data packets are immediately discarded and the data packets are not forwarded to its next hop neighboring node. It generates information "black holes" in the network, so it is called a blackhole attack, which is difficult to understand for routing nodes in the WSN. Such malicious nodes can be internal legal nodes identified by external intruders. Blockchain-based protocols can ensure detection of malicious nodes and intrusion and can access control for better overall protection of the sensor network. Blockchain takes a variety of forms, such as time stamping, distributed consent, data encryption, and economic incentives. It is used to solve problems involving inefficiency, high cost, and unsafe data storage.

6.6 Blockchain in Energy-Aware Intelligent Routing Protocols for Wireless Sensor Networks

There are some blockchain-based energy-efficient intelligent routing protocols for WSN discussed in detail below:

6.6.1 A Trusted Routing Scheme Using Blockchain and Reinforcement Learning for Wireless Sensor Networks. [5]

As a self-organizing, reliable, decentralized ledger system, blockchain is well suited for multi-hop distribution WSNs. Over the years there has been a lot of research about applying blockchain to routing algorithms. A new trustworthy routing strategy based on blockchain and reinstatement education is discussed here. Also planned based on learning to reinstall a particular routing scheduling algorithm for a planned blockchain-based network architecture, reinforcement learning and blockchain-based (RLBC) is labeled as the routing algorithm used to help. The optimal routing node is listed below. System security and fairness cannot be guaranteed, and third-party trust management centers are at risk of attack. During this time, trusted third-party administration centers are at risk of being attacked and controlled by malicious nodes, and therefore the security and integrity of the system cannot be guaranteed. Blockchain is basically an extensive data collection managed by multiple nodes and often addresses issues of trust and security. The methodology of the threat of the plan is discussed first, and then the attack and deception methods of the bad nodes in the marine environment are briefly described. Threat plan is a specific schedule based on learning reinforcement.

6.6.1.1 Threat Model

This network structure assumes that the blockchain network is reliable, which means that no attacker can control the blockchain network by controlling more than half of the nodes' servers. We also believe that route nodes are unreliable and that a malicious attacker can monitor sensitive route nodes. A malicious head node may mistakenly claim that a certain number of packets were sent to the head node during scheduling or refuse to accept packets sent from another head node.

Incorrect routing information, such as row length information, can be published by malicious routing nodes on the routing network, which affects the process of routing shedding. They can serve as nodes of blackhole attacks and refuse to forward packets. However, the conspiracy attack of two routing nodes to complete an illegal blockchain transaction is not considered. Moreover, it is assumed that a routing node can function only as a regular or malicious node, which means that the attacks are not scattered in any way. Meanwhile, sometimes abnormal behavior due to node functioning is not taken into account (for example, no node sends messages on time or loses wireless spectrum).

6.6.1.2 Blockchain-Based Network Architecture

Integrated blockchain, a distributor with tamper-proof, decentralization, and traceability features, is used to increase the reliability and visibility of routing information on wireless sensor networks and to record related information related to blockchain token transactions. Each node, as shown. The basic structure is bifurcated into two parts: the blockchain network and the original routing network. There are three types of entities in the structure: routing node R, server node S, and terminal computer. A real routing network consists of routing nodes and terminals. Packets are sent from the source terminal to the destination terminal on the routing node. Ri then selects the next hop routing node Rπ through the routing policy. After continuous activation, the packet will be delivered to the target routing node RT and then to the target terminal. Each blockchain system has a specific sensor reduction algorithm to ensure transaction fairness. They represent two entities in a network of Proof of Authority (PoA) blockchains with different identities:

- Validator: Validators are blockchain counter-authentication nodes that have received advanced approval and are responsible for POA blockchain verification work. Each server node in the system has a validation provider and a unique blockchain address in the POO blockchain with higher rights. Their special features are smart contract execution, blockchain transaction verification, and blocks on the blockchain. A new verifier with more than 50% of the vote can be added through the selection of certified verifiers. Even if a corrupt legitimacy occurs, at most it can attack only a complex block; on the other hand the votes of the legitimacy can expel the corrupt legitimacy.
- Minion: Minions are less likely nodes and cannot be validated on the PoA blockchain. Each routing node is a minion in the system. PoA blockchain has a unique address for height and blockchain.

They may want to initiate token agreements, trigger specific agreement features, and know the details of blockchain transactions. In this system, the nodes in the WSN are stable. This does not apply to inbound and outbound programs. The status of the blockchain system is updated actively. Flow nodes can initiate signal agreements to create tokens. They will exchange tokens through token agreements to send tokens based on the sending and receiving of tokens. Tokens represent packets between specific routing nodes. Unlike a traditional routing structure, each routing node is registered in the registration agreement after entering the network. Route information is passed to server nodes through the blockchain sensing process. The study pattern for each routing node will extract data from the blockchain and send it back to the routing node to the routing policy.

1. Mapping *map: blockchain.address→physical.address;*
2. Mapping *state: blockchain.address→0 or 1;*
3. **while** *true* **do** /* The contract is waiting for a contract caller to trigger */

Input: Contract Caller's Blockchain Address *ba;* Contract Caller's Physical Address *pa;*

Output: Registration Result *r;*

4. *r* ← *null;*

5. **if** *state(ba)* = 1 **then**
6. *r* ←*failure;*
7. **else**
8. *map(ba)* = *pa;*
9. *state(ba)* = 1;
10. *r* ←*success;*
11. **end if**
12. **end while**

ALGORITHM 1. PROCEDURE OF REGISTERING A NODE

Input: Environment E; Action Space A ; Initial State x0; Reward Discount γ; Learning Rate α;
Output: Policy π;
1: Qt(x,a) = 0, P(x,a) = 1/|A(x)| ;
2: x = x0;
3: for T = 1, 2, ... do
4: a = πp(x);
5: r = reward by routing action a;
6: x0 = next state by routing action a;
7: a0 = π(x0);
8: Qt(x,a) = Qt(x,a) + α(r+γQt(x0,a0) – Qt(x,a));
9: π(x) = arg maxak qt(x,ak)•Qt(x,ak);
10: x = x0;
11: end for

The blockchain-based trust model for WSN is discussed in the program. This form is used to prevent malicious attacks and to securely transmit information from a standard sensor to a node. The basic principles of this model or program are as follows:

- It highlights a trustworthy pattern to avoid retaliation.
- The smart contract is integrated using the following tools: Ganache, Remix IDE, MATLAB R2018a and MetaMask.

In Figure 6.8, the key elements for this system model are categorized as follows:

i. Ordinary sensor nodes: This project describes a blockchain-based trust model for WSNs. This model is used to avoid malicious attacks and to securely transfer data from the synchronization of data from normal sensors.
ii. Sink nodes: Three main functions are performed by these nodes. The first is to collect data from a common sensor. The second is to add new nodes using the PoA

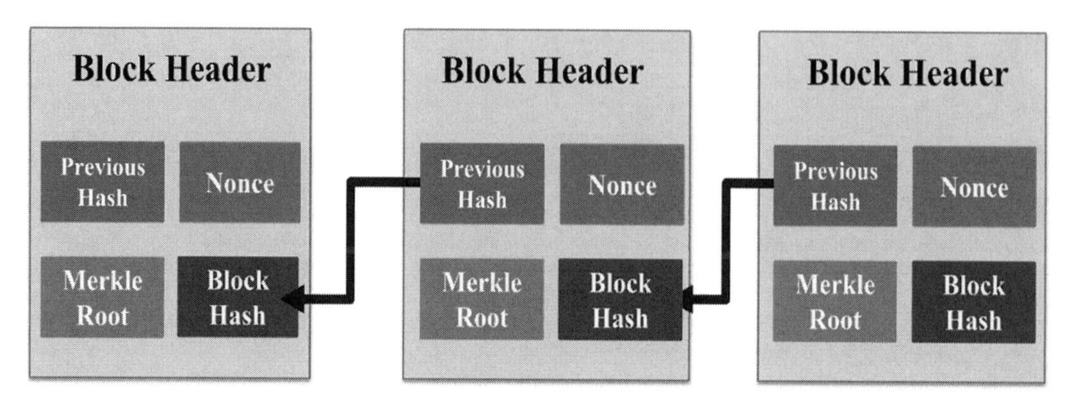

FIGURE 6.8
Representation of typical public ledger of bitcoin.

accreditation reduction process. In the third function smart contracts are issued by the main server. Data is separated into duplicate nodes based on common node ID and location. Transactions are synchronized to keep records. Each node has its own database composed of hashes. Each synchronization node is the primary server with the ability to communicate with other synchronization nodes and common sensory nodes. The synchronization node uses a private key to access information on the primary server.

iii. Main server: The main server is also known as the base station or end point. The main task of the base is to publish intelligent contracts: to operate and disseminate sensitive information. The primary server records each payment transaction with a duplicate ID and location in a static database. Only master servers or pre-authorized duplication nodes can access this database.

This can be seen in Figure 6.9. Nine common nodes are connected to the node of the sync. Data from common sensor nodes will be retrieved by any sync node. Sync nodes are able to send their data to other syncs and to the main server. A smart agreement is implemented in the sync nodes and issued by the main server. Sync nodes are able to authenticate and blacklist any common sensor node to detect malicious activity at any time.

Each sink has its own communication log and other nodes in the other distributor account. In this system model, the authenticity of the data is checked on the duplication nodes. One thing to note is that access to the master server is given only to the synchronization node. The primary server checks the operational status of the synchronization nodes and common nodes.

It can also remove a node when it dies or performs a suspicious activity. Hash function: The hash of each transaction, called the transaction hash, is a function that takes the value of the input and produces the value of the output. This output, as opposed to the input value, is the set value. It's written as

$$f(a) = b$$

where "a" is connected to an input and "b" is connected to an output opposite to "a." Hash values are usually "unchanged," which means that the input cannot be multiplied by knowing the output except for hits and test methods.

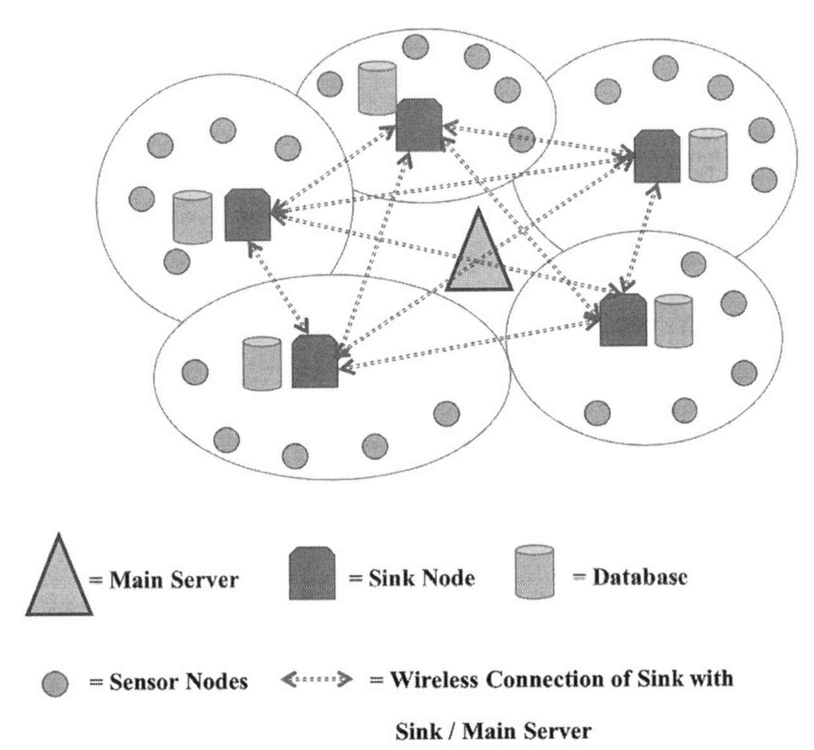

FIGURE 6.9
Blockchain-based system model.

6.6.2 A Blockchain-Based Contractual Routing (BCR) Protocol for the Internet of Things Using Smart Contracts [6]

In general, existing routing protocols have two broad stages. The first step is for root development, and the second step is for root protection. The IoT resource sends a path request (RREQ) control message in Step 1 to find the path to the target device. Each proxy or destination device that receives a RREQ packet can respond by sending a route reply (RREP) message to the IoT device. A route error message (RERR) is used to warn that a specific product could not be obtained from another device. It must delete the route from their routing table. Instead of generating an RRQ control message, the BCR protocol generates a smart contract that requires each source IoT system to go to a destination or data gateway for a specified period of time. A separate address created by the block manufacturer when a smart contract is placed on a block.

The IoT system will send this address to their neighbors to request a new route. Smart contracts within the blockchain are used to implement the BCR protocol. IoT devices requested. Therefore, the BCR protocol can perform functions in a smart contract. Therefore, the BCR protocol replaces smart contract function calls to transmit messages within the routing protocol.

I

BCR Protocol States:

6.6.2.1 *Route Requested*

The source IoT device creates a smart contract within the blockchain if it enters the gateway and sends a smart contract address to their neighbors. This path determines the state domain in the discovery agreement. The source IoT device creates a smart contract within the blockchain if it enters the gateway and sends a smart contract address to their neighbors. This path determines the state domain in the discovery agreement. IoT products do not really need to know the address of the data gateway. Instead of creating smart contracts, you can use a different IPv6 address system. The source IoT tool transfers some of its own blockchain tokens as contract addresses to smart contract addresses.

Medium IoT products that are likely to receive tokens will respond to routing requests. The source IoT device also sets a valid time for the routing request under the smart contract. This is called the smart contract initialization.

6.6.2.2 Route Offered

Any neighboring IoT device that has a secure gateway access and wants to be part of the data packet relay will respond to the domestic smart agreement and call a feature in the parent contract. By transferring some of its own tokens to a smart contract address, the proxy IoT machine delivers its services to the source machine.

Roll calls send block makers to a network that converts the received smart contract status into a route provider. Each protocol can store up to three routing offers from different IoT products. Neighboring IoT proxy systems mean a new smart agreement because they do not know the way to the gateway or destination.

The arbitrator agreement basically stores the address of the issued smart agreement or any other intermediary agreement in a parameter.

6.6.2.3 Route Accepted

The source IoT system decides whether the given route should be accepted for sending its data packets. Based on the internal principles of the entrance, it selects the next neighbor. To secure data packets and increase throughput, it can choose a low-cost route provided by its neighbor or multiple neighbors to serve as a relay.

6.6.2.4 Route Passed

Smart contract status data is converted to gateway data when it is received by the gateway. If the IoT-mediated device B provides a path that cannot successfully transmit data packets to the IoT device as specified in the smart contract for a specified period of time, the source IoT device cannot successfully forward the packet with the B restriction. Time period on your inner blacklist.

The source IoT device will add its existing blacklist addresses to the blacklist of any newly developed smart contract.

6.6.2.5 Routing Aborted

At any time, the routing process can be terminated by calling the recovery feature in each smart contract from each IoT network. However, the smart contract aort works depending on its IoT device form and the status of the current contract.

6.6.2.6 Routing Expired

Since the BCR protocol has different timers, the IoT system can request an expiration feature within the smart contract, which is to check and comply with the timer.

BCR protocol function: Called contract functions are transferred to the status of the smart contract. Each time a function is called from an IoT node, the tokens specified in the gas function will be transferred from the blockchain account of the IoT device to the block manufacturer's account.

i. Route request: Blockchain creators can request that a smart agreement be created on the blockchain if each IoT product needs to add a data gateway to the destination. For this function, the source IoT system digitizes the system and sets the terms of the smart contract. Algorithm 3 shows this function.
ii. Route offer: This occurs when an intermediate IoT device is ready to transmit data packets to a destination or data gateway in its internal routing table and to transmit data packets to the source IoT device. Each contract approves proposals for a maximum of three channels from the arbitration panel. Algorithm 4 shows this function.
iii. Route accept: When the source IoT device selects the proposed route, it performs routing within the blockchain. If the function caller's IoT device address is the same as the source IoT device in the smart contract, the block manufacturer performs this function. Algorithm 3 shows this function.

ALGORITHM 3. ROUTE REQUEST FUNCTION

1: **function** ROUTE REQUEST(DESTINATION, RRB, RRE, BLACKLIST, PARENT ADDRESS(OPTIONAL), HOP(OPTIONAL))
2: Transfer Gas tokens from the function caller to the block producer.
3: Transfer RRB tokens from the function caller to the current contract address
4: Set RRE to *Route_Request_Expiry*
5: Set Blacklist to *Blacklisted_Addresses*
6: **if** this is an original smart contract **then**
7: Set *Hop* to 0
8: **end if**
9: **if** this is an intermediary smart contract **then**
10: Set *Hop* to Hop
11: Set *Parent_Contract* to ParentAddress
12: **end if**
13: Set *Timestamp* to Now
14: **end function**

ALGORITHM 4. ROUTE OFFER FUNCTION

1: **function** ROUTE OFFER(ROB, ROV)
2: Transfer *Gas* tokens from the function caller to the block producer.
3: **if** the function caller address is not in *Blacklisted_Addresses* and the number of offers is less than three **then**
4: Transfer ROB tokens from the function caller to the current contract address
5: Set ROV to *Rout_Offer_Validity*
6: **end if**
7: **end function**

ALGORITHM 5. ROUTE OFFER FUNCTION

1: **function** ROUTE ACCEPT(INTERMEDIARY)
2: Transfer *Gas* tokens from the function caller to the block producer.
3: **if** the function caller is *Source* **then**
4: Move the intermediary to *Selected_Route*
5: Transfer the ROB tokens of the other intermediary devices back
6: **end if**
7: **end function**

6.7 Conclusion

Routing protocols play a really important role in creating fewer and more effective communication interruptions between source and destination nodes. A network's service, performance, and reliability largely rely on the choice of a great routing protocol. In heterodox situations, WSN has the ability to track, gather, and send data from one location to another. This network does, however, have a lot of security threats. As it has a solid security model backed by cryptography, blockchain technology can solve the problem of both energy conservation and security risks. The public ledger principle, which stores public records of events on the network, is very useful for identifying malicious nodes in the network and improving routing performance. For the Backpressure (BP) protocol [4], routing nodes provide reliable and active routing information on the blockchain network.

In order to choose the best routing path and avoid routing links with malicious nodes, a comprehensive reinforcement learning model is used. As a result of efficiency estimates, the BCR Protocol [5] shows that BCR reduces overhead routing by five factors related to AODV. It is immune to attacks from blackhole and greyhole as well. Therefore, for wireless sensor networks, blockchain-based energy-aware protocols can be very helpful and can be a possible solution to power-consumption problems.

References

1. N. G. Palan, B. V. Barbadekar, and S. Patil, "Low energy adaptive clustering hierarchy (LEACH) protocol: A retrospective analysis," 2017 International Conference on Inventive Systems and Control (ICISC), Coimbatore, 2017, pp. 1–12, doi: 10.1109/ICISC.2017.8068715
2. W. Guo, W. Zhang, and G. Lu, "PEGASIS Protocol in Wireless Sensor Network Based on an Improved Ant Colony Algorithm," 2010 Second International Workshop on Education Technology and Computer Science, Wuhan, 2010, pp. 64–67, doi: 10.1109/ETCS.2010.285
3. A. Manjeshwar, D. Agrawal, (2001). TEEN: A Routing Protocol for Enhanced Efficiency in Wireless Sensor Networks. Intl. Proc. of 15th Parallel and Distributed Processing Symp. 22. p. 189. doi: 10.1109/IPDPS.2001.925197
4. A. Sarkar, T. Maitra, and S. Neogy (2021) Blockchain in Healthcare System: Security Issues, Attacks and Challenges. In: Panda, S.K., Jena, A.K., Swain, S.K., Satapathy, S.C. (eds) Blockchain Technology: Applications and Challenges. Intelligent Systems Reference Library, vol 203. Springer, Cham. doi: 10.1007/978-3-030-69395-4_7
5. J. Yang, S. He, Y. Xu, L. Chen, and J. Ren, A Trusted Routing Scheme Using Blockchain and Reinforcement Learning for Wireless Sensor Networks. Sensors 2019, 19, p. 970.
6. R. Gholamreza, L. Cyril. (2018). A Blockchain-Based Contractual Routing Protocol for the Internet of Things Using Smart Contracts. Wireless Communications and Mobile Computing. 2018. pp. 1–14. doi: 10.1155/2018/4029591

7

Intelligent Vehicular Ad Hoc Networks and Their Pursuance in LTE Networks

Arjun Arora

CONTENTS

7.1 Introduction

Roadside units (RSUs) and onboard units (OBUs) are outfitted with communication modules, information processing systems, and sensors. These modules make it possible for vehicles and infrastructure units to communicate with one another over multiple hops to share and exchange information concerning the routine driving status reports of vehicles and the varied driving environments. With all these mechanisms, RSUs and the OBUs form a network, which is the first version of mobile ad hoc networks. Vehicular Ad Hoc Networks (VANETs) have various applications.

DOI: 10.1201/9781003102397-7

The major thrust behind this kind of network is software associated with traffic safety. Tens of thousands of individuals die and hundreds of thousands are injured in traffic accidents. Providing information about conflicts avert most accidents. VANETs also improve traffic optimization. These mechanisms allow transport government authorities to steer vehicles and handle them electronically (e.g., rate controller, velocity, etc.), which is a lot more efficient compared to conventional manual management.

In addition to safety-related programs, value-added services can be given through VANETs. By implementing electronic payment protocols in VANETs, an individual can expect to handle a toll collection channel. Since GPS programs have become available in many [1] vehicles, it's also possible to comprehend location-based providers in VANETs, for example, finding the nearest gas station, restaurant, resort, etc. Electronic entertainment is included with other sorts of services and so forth. These services result in an unpleasant driving experience for most motorists. For all solutions mentioned in published works to make life easy, they ought to rely on protected and privacy-preserving protocols that encourage users to engage without fear for their safety or privacy. Unless appropriate steps are taken, numerous attacks may easily occur, namely message modification, identity theft, false information production and propagation, etc. It will become clear that message authentication, integrity, and non-repudiation are requirements in VANETs. There's a need for mechanisms offering VANETs to use security(i.e., protocols, methods, and processes that are ready to detect if a message was altered by an attacker and also determine who is the actual sender of a message [2, 3]).

Apart from these vital security conditions, privacy is another problem in VANETs that can't be forgotten. If privacy protection measures are jeopardized, the privacy of VANET consumers may be compromised. For instance, an eavesdropper may gather messages delivered by vehicles and monitor their locations; they can extract users' information such as identities and homes by doing this. Be aware that these privacy issues are much like location-based services [4, 5]. Yet, privacy in VANETs ought to be automatic. This can be user-related details like a license plate, current speed, current location, identification number, and so on, which ought to be kept confidential from other users/vehicles while authorized users must have access to it.

7.2 Overview of Vehicular Ad Hoc Networks

Today's latest automobiles have an intra-vehicular network that enables wireless communication between automobiles and electronics such as smart mobile, global positioning system (GPS), and Bluetooth networking players. However, an inter-vehicular communication network is not accessible. Vehicular ad hoc networks (VANETs) are described as a subset of mobile ad hoc networks (MANETs). So node (i.e., automobile motion) is limited by road training circuit, surrounding traffic, and traffic regulations. Due to these limitations, VANET is supported by a few infrastructures that help with some professional services and give access to networks. The infrastructures are set up at areas such as roadsides, service channel intersections, or areas using weather conditions. Figure 7.1 explains how vehicles communicate with one another and with fixed infrastructure.

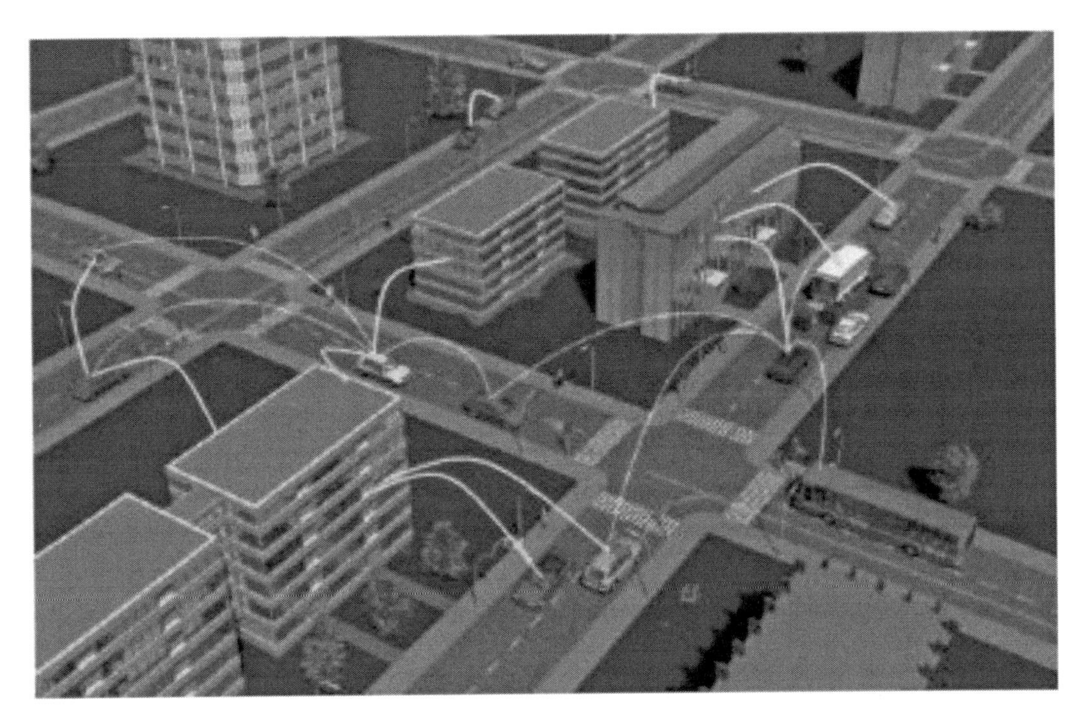

FIGURE 7.1
Vehicular ad hoc networks.

7.3 Characteristics of VANETs

7.3.1 High-Dynamic Topology

VANET has a dynamic topology. The communication connections between nodes change quickly. Communication between two nodes lasts for a short time. For instance, if two vehicles moving off from each other at a rate of 25 m/s and the transmission range is approximately 250 m, then the connection is going to last for just 5 seconds (250 m/50 ms^{-1}). So this how dynamic topology is present in VANET.

7.3.2 Frequent Disconnected Network

In the above characteristic, we can see that the connection between a couple of vehicles remains for 5 seconds. To maintain constant connectivity, vehicles require another connection nearby instantly. But if failure happens vehicles can connect with a roadside unit. Frequent disconnected network chiefly occurs where automobile density is low, such as in rural places.

7.3.3 Modelling and Prediction

The above two attributes for connectivity require the knowledge of the location of vehicles and their movement, but this is difficult to predict since automobile movement doesn't have a pattern. So models of node prediction based on the study of predefined road roadways and vehicle speeds are used.

7.3.4 Communication Environment

The mobility model varies from highways to town setting, in different surroundings. So modelling and automobile movement prediction and algorithms must adapt to such changes. Highway models are easy because automobile movement is one-dimensional. However, in the case of town surroundings, a lot of vehicle barriers such as construction exist, making communication programs complicated in VANET.

7.3.5 Hard Delay Constraints

In case of safety, application delay constraint is important in scenarios such as injury in case of sudden breaking and other crisis situations. VANET program depends upon data communication time. It can't compromise for delay in information in the case of high speed VANETs.

7.3.6 Interaction with On-Board Sensors

The on-board detector in a motor vehicle can be utilized to identify vehicle location and vehicle motion. These data and vehicle speed are utilized for communication between vehicles.

7.3.7 Mitigating Features

It follows that security of vehicular networks faces a large number of challenges. But we see that VANETs have characteristics that could mitigate the challenges and allow security and privacy standards. Unlike in the majority of mobile ad hoc networks (MANETs), nodes in VANETs could be anticipated to have substantial energy sources and computational capability [6]. Automobiles have a considerable supply of electricity when compared with detectors or cell phones. This suggests that the communication protocols don't have to be continuously active. Compared with a car's purchase price, OBUs' price isn't a problem; actually, an OBU could be presumed to be as successful as a computer. The protocols can exploit innovative cryptosystems to attain security in VANETs, in comparison to other mobile networks, in which minimization of all operations is essential [7, 8].

Security protocols in VANETs may gain from the transport systems. In the majority of nations, all vehicles have to be registered in a central authority, making vehicle identification possible, by way of example. Traffic lights, traffic sensors, radar, etc., are part of the infrastructure of current transportation systems. They may be upgraded to become roadside units in VANETs. Additionally, vehicles undergo routine (yearly) safety inspections that allow sanity checks to be conducted against the elements of the vehicular networking method of every vehicle. Tamper-proof OBUs could be assessed for integrity and upgraded to the producer's most up-to-date version. Malfunctioning detectors that offer false data (possibly due to tampering by an adversary) could be substituted. Additionally, vehicles can leverage the input derived with information offered from the networking subsystem. In most situations a driver can perform a much better appraisal of a scenario and reliably assess whether the information is right or not and if the crucial information is individually identifiable.

Redundancy in vehicular communications can be an element that helps to enhance security and alleviate the burden of message validation in VANETs. A car receives large quantities of messages. A number of them might be reporting traffic requirements; this may be used to fix messages brought on by attacks or by mistakes [9–11]. Further, just a small percent of these messages require validation. By way of example, a notification that the automobile will speed up doesn't influence the vehicle. A car doesn't have to validate a notification. By taking these factors into consideration, the message validation and waiting time loss could be relieved without security.

Lastly, law enforcement mechanisms are most likely to show reluctant behavior in vehicular networks, which endangers motorists' safety. This may be a hindrance to the users of VANETs. Be aware that law enforcement isn't available in different kinds of wireless networks. To exploit this specific deterrent, vehicular communications have to provide non-repudiation so the message generator can't deny the simple fact they created a message. In the event of a dispute, messages could be taken as proof.

7.3.8 Applications of VANETs

The aim of VANETs is primarily to enhance safety on the road. To make this happen, the vehicles behave as detectors and exchange messages with other vehicles. These messages contain information such as the speed of automobiles, condition of the road, and traffic density. This permits governments and drivers to respond early to some situations that are dangerous, such as accidents and traffic jams. Nevertheless, the study within the subject of VANET has found programs and technologies.

7.3.8.1 Type-1: Programs for Safe Navigation

This program handles various important terms of traffic safety. Many solutions can be provided, including the following:

- Program for preventing collision by using distance calculation between two vehicle to use abrupt braking method.
- Program for recognition of dangerous and damaging driving circumstances. These circumstances could be a damaged road, obstructed road, or road covered with snow or mud.
- Program for emergency call providers after an accident happens. Here the vehicle can immediately call an expert if an accident happens.
- Programs for finding rogue drivers who are disobeying traffic guidelines like exceeding the speed limit, chatting on the phone while driving, or driving on the wrong part of the road.

7.3.8.2 Type-2: Programs for Traffic Regulation and Internet Connectivity

The second type of programs are the following:

- Program for advanced navigation aid (ANA) such as car park formation, real-time vehicle congestion information, anticipated weather condition, etc.,
- Internet connection providers can be supplied to vehicles for travel convenience and enhanced efficiency. This is prepared by data transfer between vehicles and roadside units.
- Chatting solutions between users of the same root. This can enhance driving safety because one driver can send an instant caution message to other drivers
- Program for advertising local/closest service stations, hotels, shops, and malls.

7.3.9 Challenges in VANETs

There are lots of difficulties that are experienced by VANET. These obstacles are technical challenges and also business challenges. We will notice these difficulties in the following paragraphs.

7.3.9.1 *Technical Challenges*

There is no transport protocol for vehicular networks so far, and it's very complicated to design protocols or to accommodate present protocols by modifications. The transportation of messages between nodes should be completely free because there will be no time to retransmit the message [12]. These messages are essential, so transmission should happen instantly, and should be error-free and read correctly over a period of time. Most frequently, transport messages have to be error-free because a message won't be able to be retransmitted in a case in which two cars are passing each other while moving in opposite directions; the two cars may be out of connection range for a while. A VANET might need to boost its transport layer protocols from protocols in ad hoc network transport technologies. We have to use a protocol designed for VANET other than Transmission Control Protocol [13, 14]. Recent challenges that need to be overcome to create a transport protocol are long round-trip times, high packet-loss rates, high probability of packet reordering, and connection durations that exist in current ad hoc networks. Vehicular ad-hoc networks add to the complexity because the nodes are travelling at high rates of speed. In general, VANETs must work in all kinds of traffic (i.e., high and very low vehicle density environments in urban and rural surroundings respectively) [15, 16]. This creates a challenge for the hardware design for VANETs. Because, for example, in a low-density vehicle environment the number of vehicles will be less, so some vehicles are going to be out of range for communication. In a high-density vehicle environment sharing of bandwidth is still a struggle for VANET.

7.3.9.2 *Business Challenges*

The business challenge for VANET would essentially be to make a new sector. Additionally, a client may not realize all that is possible with VENET, so advertising and awareness will have to be increased. One of the business challenges for VANETs is that there will probably be vehicles on the road that are not outfitted with VANET.

7.4 Vehicular Ad Hoc Networks

Conventional traffic management techniques typically relate to a centralized infrastructure by which sensors and cameras installed on the road gather details about the density and traffic conditions. These details are delivered to a central unit to process and make good decisions. Such programs have a cost of installation and are distinguished by a response time of one minute for the processing and transfer of information's purchase. In a circumstance where the deadline is of significance and transmission of data is critical, this delay is unacceptable. Additionally, the gear installed on the road demands upkeep [17]. To be able to deploy a scale, considerable funding in communication infrastructure and detectors is important. However, with the rapid development of wireless communication technologies, monitoring

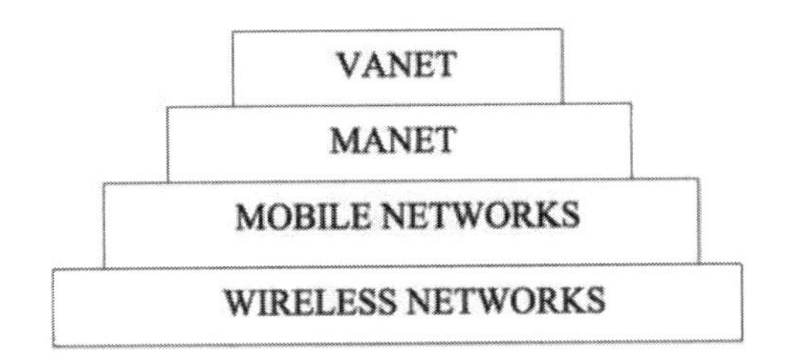

FIGURE 7.2
Wireless network hierarchy.

systems, and data sets by detectors, a brand new decentralized (or semi-centralized) architecture predicated on vehicle-to-vehicle communications (V2V) was designed in the past several years attracting great interest from the scientific community, automobile providers, and telecom operators [8, 9]. This sort of architecture relies on a dispersed system and can be shaped by the vehicles without the support of a fixed infrastructure relaying info and messages.

A VANET network is a characteristic of MANET networks in which mobile nodes have been vehicles (smart) outfitted with computers, network cards, and detectors. In the same way as any other ad hoc network, vehicles may communicate with each other (such as to swap traffic info) or use base stations placed along the roads (to request advice or access to the internet). Figure 7.2 shows the hierarchy of wireless networks in which the addition of VANET from MANET, MANET networks at the mobile and mobile networks in wireless networks, can be seen. The debut of intelligence within the sphere of automobiles is directed at enhancing the lives of drivers and passengers. The software is innumerable and varies from safety and relaxation through solutions and entertainment. These ideas would be of great interest in what is often called intelligent transport systems (ITSs). The idea would be to provide intelligence in vehicles by equipping them. At that level it is referred to as local board intellect (providing only neighborhood vision and the surrounding environment). Vehicular networks comprise two sorts of software that construct an ITS software involving individuals or may be even to trigger in case of any passenger and or driver is frightened [1, 4]. Vehicular networks will become transport systems' cornerstone. From an architectural perspective, communication in a VANET may be vehicle-to-vehicle (V2V), vehicle-to-infrastructure (V2I), or hybrid.

7.4.1 Vehicle-to-Vehicle Communication (V2V)

In this class, a vehicle network sometimes appears as a distinctive instance of MANET in which electricity constraints and memory capacity are relaxed as well as in that the model is not arbitrary but predictable with freedom. This architecture may be found in diverse situations occurrence of alarms (emergency braking, collision, lag, etc.) as shown in Fig. 7.3. In case no VANET substructure is utilized, then no standard setup is important on the roads; and vehicles are made to keep in touch with one another anywhere, whether highways, mountain roads, or city streets, which could offer an even more flexible and much less costly communication. This strategy suffers from certain drawbacks:

- The period of communication is large, given the communication is completed using multi-jumps.
- There are frequent disconnections because vehicles are mobile.
- Network security is quite restricted.

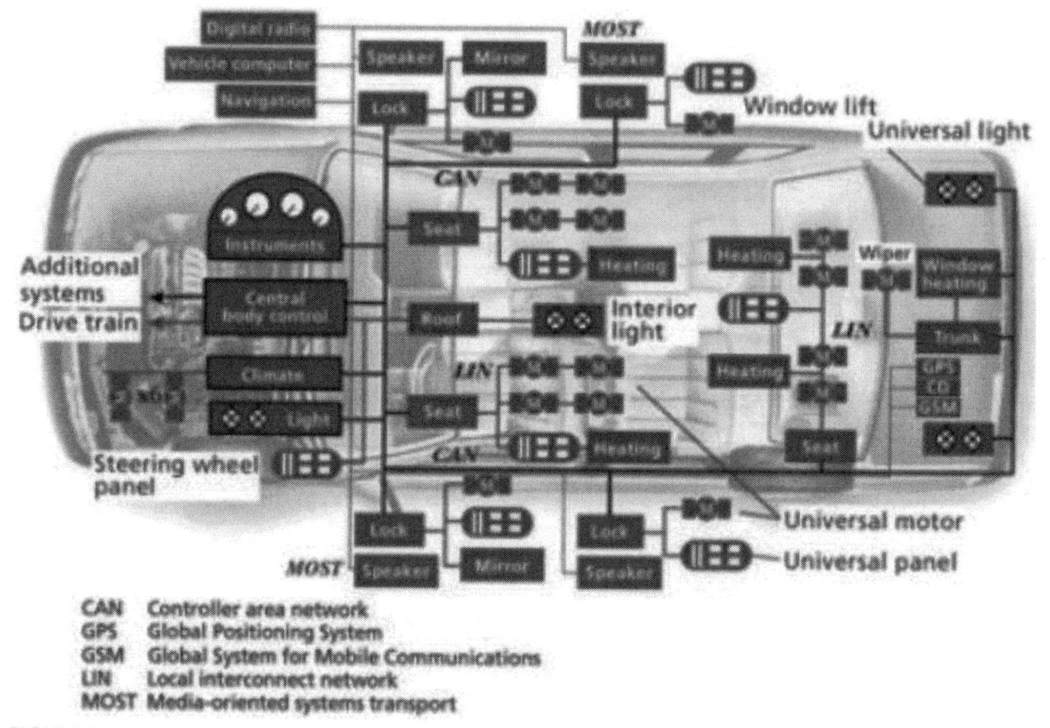

FIGURE 7.3
Network architecture design of a modern vehicle.

7.4.2 Vehicle Communication with the Use of Infrastructure (V2I)

In this classification, we don't simply concentrate on easy inter-vehicle communication methods but additionally on those using base stations or points of RSUs (name explained because of the consortium of Car 2 Car communication). This plan is based on the client/server model in which vehicles are customers and stations installed along the road will be servers. These servers are attached through a wired or wireless interface [3]. All communication has to proceed through them. They could offer users more services on trafficking, connect to the internet, exchange car-to-home communication information, and direct the car into the garage for the diagnosis. The main disadvantage of the approach is the fact of costly and time-intensive setup with these stations along the roads as well as the expense of maintenance of these channels.

7.4.3 Hybrid Communication

The mixture of both types of communication architecture gives a fascinating hybrid architecture. The development of infrastructure is fixed, using vehicles since relays can expand this space. Nonetheless, inter-vehicular communication has problems with routing issues. In instances like this, access to infrastructure may enhance network performance. To be able to prevent multiple base stations being used by vehicles, intermediate hops are required and are of utmost importance. An alternative case for this hybrid architecture is the vehicular sensor network (VSN). This sort of network functions as a completely new automobile network architecture because automobiles are provided with more detectors of all types (pollution detectors, rain sensors, tire condition detectors, Electronic Stability

Program, Anti-Lock Braking System, satellite geolocation, etc.). Information supplied by the unit may help get information on road traffic (congestion, delays, average traffic rate, etc.) and parking areas and more general information, such as average fuel consumption and the speed of contamination, or it may be used for tracking programs (via cameras on the vehicles).

7.4.4 Pursuance of VANETs in LTE

In order to assess the pursuance of VANETs in long-term evolution (LTE) networks, we need to start with modelling and analysis of the queuing delay in a highway scenario of a VANET network with V2V discovery. Queuing delay is simply the time required by a vehicle that is not in range to come into range of a given RSU. A highway scenario is considered to be multiple lanes with each lane having its own speed limit. For a given lane "k," the arrival time is distributed exponentially, having traffic density of λ_k [15, 16]. When having low to medium traffic flow it is called L_{P_k}. In each lane "k" is considered without any loss as described in [10, 17]; the vehicles are assumed to have multiple speeds that are distributed randomly with a normal distribution between maximum and minimum of speeds at which these vehicles are driving in sync with other measurements [13]. The space between two or more vehicles, denoted as "M" on a given lane "k," is exponentially distributed, having the following parameters $\lambda_{m_k} = \lambda_k / \eta_k$, where η_k is the average speed in the lane, which is equal to $(V_k^c + V_k^d)/2$ where V_k^c is the minimum and V_k^d is the maximum for speed in a given lane "k." This would result in cluster formation in each lane where two or more vehicles are in range. The research done in [16] is extended in order to support multiple lanes by means of mathematics, explaining all the characteristics of each cluster formed in each lane in a given VANET and taking into consideration the possibility of being either the lead or the last vehicle in a given cluster and considering DF (density function) for inter- and intra-cluster spacing with moderate size and moderate length of a cluster.

The concept discussed can also be implemented for a similar scenario with a two-way highway. In order to calculate DF, a cluster forms of all the vehicles need to be in range (r). Then the distance between any given set of vehicles in the same cluster would be less than "r." Furthermore the inter-vehicle distance (M) has exponential distribution [16]. The DF of intra-cluster spacing is provided as

$$E_{M_k,intra\ (M_k,intra)} = F_R[M_k \mid M_k \le r] = \frac{\lambda_{M_k} e^{-\lambda_{M_k} M_k,intra}}{1 - e^{-\lambda_{M_k} r}} \tag{7.1}$$

Furthermore the intervals between the automobile at the end of the lead cluster and the automobile at the start of a next cluster (i.e., merely the inter-cluster spacing) must have some significance more than "r." Determining if "M" is exponentially dispersed, the DF of inter-cluster spacing is provided as

$$E_{M_k,inter\ (M_k,inter)} = F_R[M_k \mid M_k > r] = \lambda_{M_k} e^{-\lambda_{M_k} M_k,inter\ -r} \tag{7.2}$$

The likelihood of an automobile being either at the start or at end in a cluster for a given lane k is denoted by "P_{N_k}." A vehicle seeking any service from any given RSU can create messages and they it could be dispersed within a given cluster in order to decrease delay in attaining an RSU. However, the message created cannot be dispersed across the cluster boundary until the first and last automobile in a given cluster are in range of the perimeter. That is why it is crucial to estimate the likelihood of becoming the first or last vehicle. The

P_N is characterized as the likelihood of existence of area wherein there are no leading automobiles as well as no last automobiles inside the range (r) for an RSU. To estimate the likelihood of being either first or last in a provided cluster in a given lane k is provided by P_{N_k}.

$$P_{N_k} = F_R\{M_k > r\} = 1 - G_{M_k}(M_k) = e^{-\lambda_{M_k r}} \tag{7.3}$$

The length between first and last vehicle is termed as cluster length and is denoted by "H_{N_k}." It is a function of cluster member (t) and $M_{k,intra}$ given as

$$H_{N_k} = \sum_{p=1}^{t-1} (M_{k,intra})_p \tag{7.4}$$

with mass function (MF) given by [16]:

$$f_q(t) = P_{N_k}(1 - P_{N_k})^{t-1} \tag{7.5}$$

As per explanation of law of total probability (TP), the DF for cluster H_{N_k} in a given lane k can be given as

$$f_{H_{N_k}}(H_{N_k}) = \begin{cases} P_{N_k} & ,t = 1 \\ \sum_{\ell \in range\ of\ t} P(t = \ell) f_{H_{N_k}}(H_{N_k} \mid t = \ell) & ,t > 1 \end{cases} \tag{7.6}$$

Equation 7.4 mentioned above shows the intra-cluster gaps in totality similar to the DF for a given cluster length. In case of a said lane the following holds true: $f_{H_{N_k}}(H_{N_k} \mid t = i) = f_{M_{k,intra}(1) + M_{k,intra}(2) + M_{k,intra}(3) + \cdots + M_{k,intra}(i-1)}\left(\sum_{p=1}^{i-1}(M_{k,intra})_p\right)$ will be the DF in totality of rapid dispersion having equivalent density and individuality. The DF in generality of this manner is shown in [13] and counts on the DF of intra-cluster space, which is computed in Equation 7.1. Equation 7.6 can be re-written as

$$f_{H_{N_k}}(H_{N_k}) = \begin{cases} P_{N_k} & ,t = 1 \\ \sum_{\ell \in range\ of\ t} \begin{cases} P(t = \ell) X (1 - e^{-\lambda_{M} r})^{-(\ell - 1)} \\ X \left(\dfrac{\lambda_{M_k}^{\ell-1} M_{k,intra}^{\ell-2}\ e^{-\lambda_{M_k} M_{k,intra}}}{(\ell - 1)!} \right) \end{cases} & ,t > 1 \end{cases} \tag{7.7}$$

By means of statistical interpolation and Monte Carlo fitting we can estimate the value of $f_{H_{N_k}}(H_{N_k})$ as

$$f_{H_{N_k}}(H_{N_k}) = \begin{cases} P_{N_k} & ,H_{N_k} = 0 \\ \alpha e^{-\beta H_{N_k}} & ,H_{N_k} > 0 \end{cases} \tag{7.8}$$

Where $\alpha = (1 - P_{N_k})^2\ x \frac{P_{N_k} \lambda_{M_k}}{1 - P_{N_k}(1 + r\lambda_M)}$ and $\beta = \frac{\alpha}{1 - P_{N_k}}$

The above estimate has a confidence proportion of 0.997. An evaluation of Equation 7.7 and 7.8 is mentioned in [14]. This estimate is easy to apply while examining delay outcomes. The equivalent estimate can be used to determine average cluster length on a provided lane k as

$$V[H_{N_k}]=\left(\frac{1}{P_{N_k}}-1\right)\left(\frac{1}{\lambda_{M_k}}-\frac{re^{-\lambda_{M_k}r}}{1-e^{-\lambda_{M_k}r}}\right) \tag{7.9}$$

The DF for a given cluster size may be utilized to acquire average delay in queuing on a given highway having several associated RSUs with different range cover. In a highway situation wherein RSUs are dispersed uniformly with a distance of W_{RSU} and with each RSU range connected at X_i at any provided point in time, an automobile cannot follow an RSU until it comes inside the range of the RSU network. To actually reduce the time needed to reach a given RSU, the automobile will have to try and reach networking RSUs via multi-hop interactions. The vehicle that is travelling in a given lane k might not be ready to send messages to any other adjoining-lane automobiles until it is in range. On the grounds of the observation the subsequent cases are explained:

Case 1: A automobile is considered to be associated if it is in range or is aspect of some cluster with a minimum of one vehicle in that cluster associated to the RSU in range. Assuming appropriate routing is now being applied in such a situation where the typical time to match a given RSU would be equivalent to zero.

Case 2: A automobile is taken as disjoint if it is placed in a region of the lane that is not in range. Essentially it must be in the region having length $W_{RSU}-2X_i$. Moreover it is presumed that on average the vehicle is situated in the center (i.e., equidistant from two RSUs) [15, 17]. Given that a vehicle is disjoint, it's going to have some common time to meet a given RSU in a given lane k in time to navigate where no RSU range is available then in that case merely half of the total length is considered and is provided by

$$E[Y_{i,k} \mid \lambda_d] \tag{7.10}$$

Where Z_k is speed of a given vehicle in lane k.

The likelihood of an isolated automobile being disjoint from an RSU network is provided by ($P_a[\gamma_d]$) and at any provided point is similarly proportional of the highway that is not in range of the RSU.

$$P_a[\gamma_d]=\frac{W_{RSU}-2X_i}{W_{RSU}} \tag{7.11}$$

The above Equation 7.11 brings out the common time in order to meet any provided RSU for any given isolated vehicle on a provided lane k to be as

$$E[Y_{i,k}]=E[J_{V,k} \mid \gamma_d]X\ P_a[\gamma_d]=\frac{(W_{RSU}-2X_i)^2}{2W_{RSU}\ .Z_k} \tag{7.12}$$

Case 3: When none of the automobiles in a given cluster are in range of a given RSU, the cluster is regarded to be unconnected. Likewise the automobiles at the boundary of an unconnected cluster are regarded to be in an exposed region $[X_i;W_{RSU}-X_i]$, and the complete duration for a dis-joint cluster could be less in contrast to the total length of the undiscovered part (i.e., $H_{N_k}<W_{RSU}-X_i$). For reference, if we take the center of the cluster then it needs to be in the area of length given as $\left((W_{RSU}-2X_i)-E(H_{N_k} \mid H_{N_k}<W_{RSU}-2X_i)\right)$.

The greatest possible length of a given disjoint cluster is $E(H_{N_k} \mid H_{N_k} < W_{RSU} - 2X_i)$ and is revealed in [12]. The possibility to find automobiles on the edge in a given cluster in a given lane k in case of exposed region is

$$P\left[J_{d_k}\right] = \frac{(W_{RSU} - 2X_i) - E[H_{N_k} \mid H_{N_k} < W_{RSU} - 2X_i]}{W_{RSU}} \tag{7.13}$$

Looking at it mathematically, we can assume the center of the cluster to generally be at the mid-point of an unnoticed region. Consequently the mean lag time to arrive at any given RSU for a given disjoint cluster vehicle travelling in lane k could be

$$E[Y_{C,k} \mid J_{d_k} \cap (H_{N_k} < W_{RSU} - 2X_i)] = \frac{(W_{RSU} - 2X_i) - E[H_{N_k} \mid H_{N_k} < W_{RSU} - 2X_i]}{2Z_k} \tag{7.14}$$

Then the mean delay for a given disjoint cluster for a provided lane k would be

$$E[Y_{C,k}] = E[Y_{C,k} \mid J_{d_k} \cap (H_{N_k} < W_{RSU} - 2X_i)]\ X\ P[J_{d_k}]\ X\ P[H_{N_k} < W_{RSU} - 2X_i] \tag{7.15}$$

Finally the total mean delay for any lane could be

$$E[Y_k] = \mathrm{E}[Y_{V,k}].P_a[n=1] + E[Y_{C,k}].P_a[n>1] = \frac{1}{2Z_k W_{RSU}}\left(P_{N_k}.A + \left(1 - P_{N_k}\right)B.C\right) \tag{7.16}$$

Where $\mathrm{A} = (W_{RSU} - 2X_i)^2, \mathrm{B} = ((W_{RSU} - 2X_i) - E[H_N \mid H_N < W_{RSU} - 2X_i)^2,$

$$\mathrm{C} = \left(1 - \frac{1 - P_{N_k}}{e^{\beta(W_{RSU} - 2X_i)}}\right)$$

and $P_a\,[n=1]$, $P_a[n>1]$ reveals the possibility of choosing remote and cluster vehicle in a given situation. The mean delay in offering services in a RSU network when vehicle is moving in multi-lane highway can also be determined.

At this point we know that the common delay for an interconnected automobile equals zero; logical models may be created using this idea, which can simplify lane delay on the presumption that the automobile looking for such a support from a provided RSU network is disconnected from the network. The worst-case situation occurs when an automobile is not able to communicate messages to a provided cluster with high quick response time. The time needed for an interaction with an RSU is merely the ingredient of two delay parameters:

a. The time until a given automobile comes within range of an automobile on a given adjoining lane.
b. The overall time required by a relay automobile to satisfy a RSU in a given network. To be able to accomplish non-zero sum the relay automobile has to be disjoint from the given RSU network.

Calculation of the estimated time to meet a relay cluster automobile in a given lane k is expected to be in center of inter-cluster gap in a provided lane "ℓ"($S_{\ell,inter}$). This assumption is rationally appropriate and mentioned in [16, 17]. Then ($S_{\ell,inter}$) has to be more than 2X and temporal delay is provided as

$$\mathrm{E}\left[T_{\ell,temp_{11}}\right] = \frac{0.5E\left[S_{\ell,inter} \middle| S_{\ell,inter} > 2X\right] - X}{Z_k + Z_\ell} = \frac{1}{\lambda_{\ell,s}(Z_k + Z_\ell)} \tag{7.17}$$

Spatial delay is comparative to average delay so as to meet RSU on a provided lane "ℓ" (E[T_ℓ]). Mean delay in delivery of content to a given RSU that makes use of a vehicle on an adjoining lane "ℓ" is

$$\mathrm{E}[T_{\ell,r11}] = \left(E[T_\ell] + E[T_{\ell,temp_{11}}]\right) \; X \; T_{\ell,r11} \; X \; P_{J_{dl}} \tag{7.18}$$

In a best-case scenario the possibility of a relay automobile in a given lane ℓ if unconnected is (P_{J_d}). The mean delay in message distribution to a provided RSU in a provided lane ℓ when the automobile is forwarding messages in k is

$$E[T_{k,r21}] = E[T_\ell] \; X \; P_{k,r21} \; X \; P_{J_{dl}} \tag{7.19}$$

In case there is a clustered client that belongs to a disjoint cluster on a given lane k that has a few relay automobile in a provided lane ℓ that is in range, then the likelihood of occurrence of such a situation would be $P_{\ell,r22}$, mentioned in [8]. In such instances delay in dispensing messages to a provided RSU equals $E[T_\ell]$, which is mentioned in Equation 7.19, pointed out earlier in the chapter. Mean delay in providing a given message to a provided RSU, having vehicle travelling in given lane ℓ when the automobile delivering messages is in lane k is

$$E[T_{k,r22}] = E[T_l] \; X \; P_{l,r22} \; X \; P_{J_{dl}} \tag{7.20}$$

Subsequently the typical queuing time needed by a message to be obtainable to a provided RSU is

$$E[V] = \sum_{k\varepsilon N_l} \sum_{i\varepsilon N/k} \{\{P_a[n=1](E[T_{l,r33}] + ET_{l,r43}] + P_a[n>1](E[T_{l,r34}] + E[T_{l,r44}]))\} \; X \; P_a[Z_k] \tag{7.21}$$

where N is set of lanes on a given road.

Contention delay happens in a wireless channel whenever there is competition among automobiles for accessibility to the channel. This study is explained in [10], where the type of delay is confirmed mathematically and provided as

$$E[h] = \begin{cases} E[h\omega].T_{st} & if \; E[h\omega].T_{st} < T_{SCH} \\ E[Q\alpha hq] + E[h\omega].T_{st} & if \; E[h\omega].T_{st} > T_{SCH} \end{cases} \tag{7.22}$$

Here $E[h\omega]$ is average contention, with T_{st} being average duration and $E[Q\alpha hq]$ being average buffering time till next scanning starts. This equals

$$E[h\omega] = \frac{h\omega_{max} - 1}{2} \tag{7.23}$$

$$T_{st} = P_{id,\beta} + T_{SU} P_{SU} + T_{CO} P_{CO} \tag{7.24}$$

$$E[Q\alpha hq] = T_{gd} + \frac{T_{CCH}}{2} \tag{7.25}$$

Here P_{id} is the possibility of a channel becoming idle, P_{su} the possibility of slot being utilized with positive interaction, P_{co} the possibility of a collision, β the length of a slot being empty, T_{su} the time needed for transmission happening, T_{co} the average time for collision, T_{gd} the length of time for guarding, and T_{CCH} the length of interval in Control Channel (CCH). It will progress the demand and paired content will be submitted to eNodeB to be able to designate needed resources.

The outcomes have been described utilizing the logical model mentioned previously. The outcomes have been shown in the form of plots. The evaluation has been done using NS3 and Monte Carlo simulation [15]. A 25-km extent of clean multi-lane highway is presumed; each lane in the highway has a separate speed. An RSU network is shown with RSUs placed at fixed ranges ($W_{RSU} = 1km$), and automobiles come in on each lane in an individual fashion following the Poisson process. An important aspect to be observed is that mobility traces used in a wide range of simulations are made on real proportions; the data given is verified and used arbitrarily [13].

All automobiles have an 802.11p unit, which has periodical accessibility. The time interval between CCH and Service Channel (SCH) is stabilized at 50 ms, with a period guard staying set at 4ms. At a given point in time RSUs have two 802.11p units with extended accessibility, with CCH and SCH providing the radio range of vehicles and RSU set at 280m following FCC guidelines. Communication among vehicles occurs in the following fashion. First vehicles are situated at W_{RSU} distance, and then one vehicle tries to associate with the RSU network randomly chosen on W_{RSU}. Figure 7.4 represents the possibility of an automobile being at the beginning on a given lane $k\left(P_{N_k}\right)$. As can be noticed in the plot, lower the traffic density improves the possibility of being the automobile at the beginning on any given lane. Figures 7.5 and Figure 7.6 show inter and intra spacing in each cluster on any given lane. In contrast with leading automobiles, as the denseness of traffic increases, we can see a decline in both intra and inter clusters.

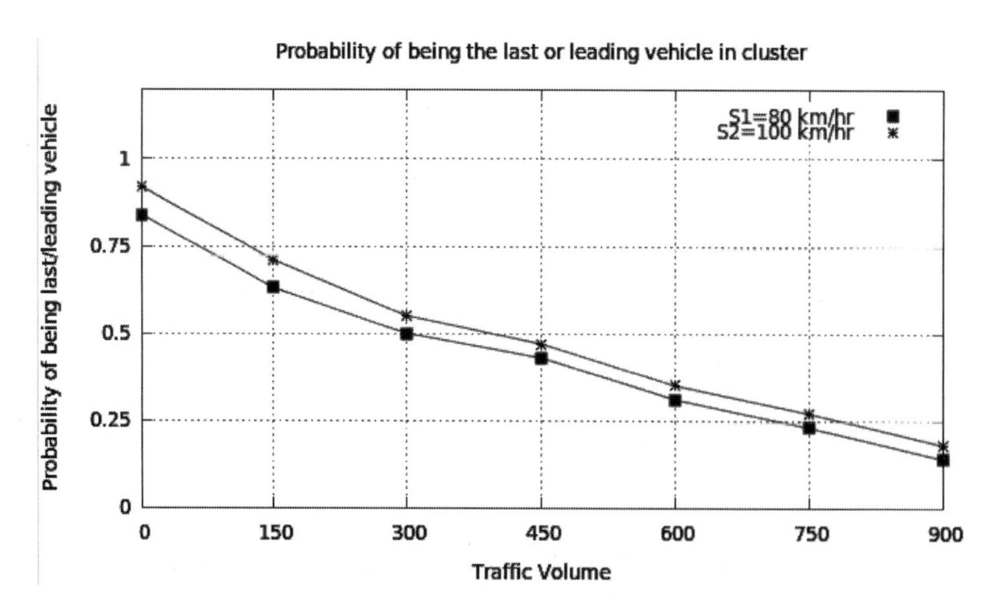

FIGURE 7.4
Probability of leading or last vehicle in cluster.

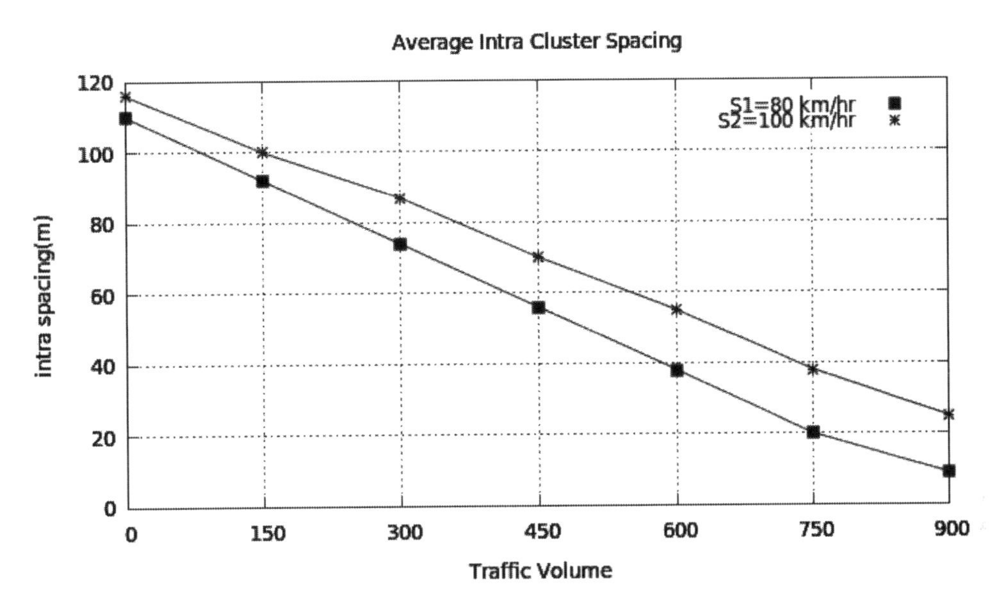

FIGURE 7.5
Average intra-cluster spacing.

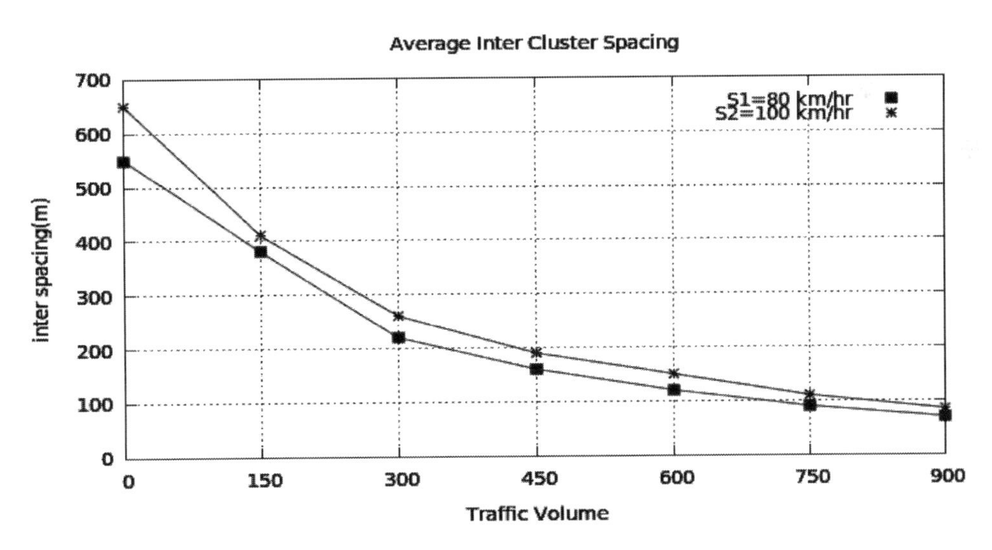

FIGURE 7.6
Average inter-cluster spacing.

Figure 7.7 shows the mean cluster length presented in Equation 7.9. As presented in this figure, mean cluster size increases with density of traffic, but it will decrease with increase in speed. Figure 7.8 shows a contrast of common delay calculated using the established model in Equation 7.23 and the analytical model [11]. An assumption is made for a three-lane highway with mean speed of 65m/s. Whenever traffic density increases, common delay decreases so that when traffic density increases the space of the cluster also increases. Figure 7.9 shows the effect of increase in number of lanes [12] with emphasis on VANET scenario. With an increase in the number of lanes the efficiency decreases

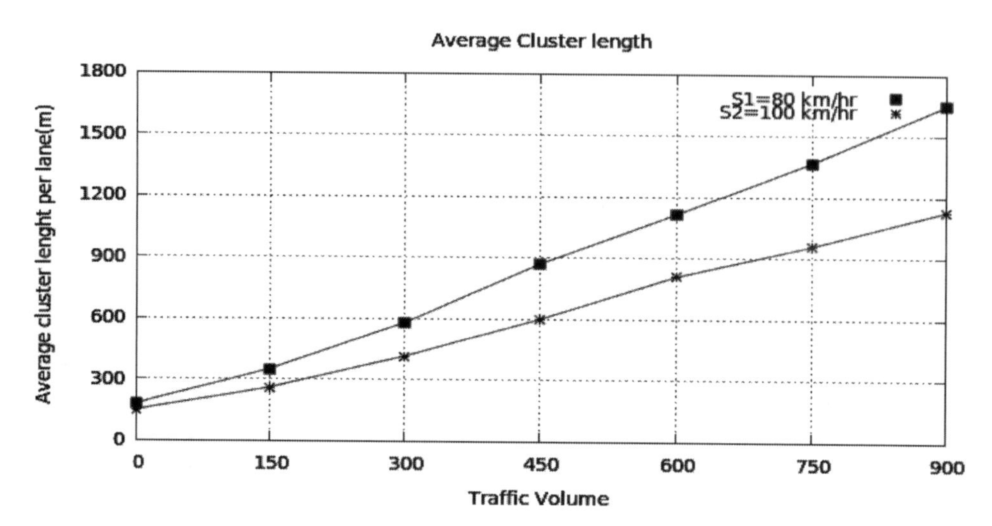

FIGURE 7.7
Average cluster length.

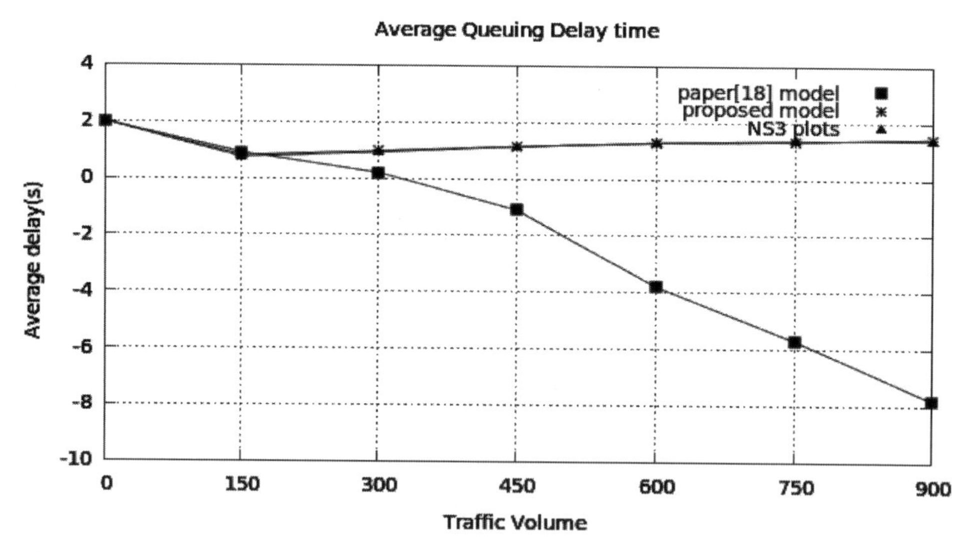

FIGURE 7.8
Average queuing delay time.

manifolds. Figure 7.9 also reveals a shortage in percentage, showing an error between NS3 analysis and the model used in the paper [16].

Figure 7.10 shows the effectiveness of routing; it can easily be observed that the typical number of packets needed to reach a given RSU in a given network is nearly constant in case of a unicast strategy. Figure 7.10 reveals the typical number of packages sent, which are then accepted correctly by common receivers till a preliminary copy is not accepted of the content by the RSU network. Figure 7.11 reveals the similarity in outcomes between a contention model and simulation outcomes. To be able to show a delay in solution, a 35-ms delay in process is included at RSU and eNodeB [14]. Figure 7.12 presents a total delay in mean in the case of the discovered protocol. Delay in the

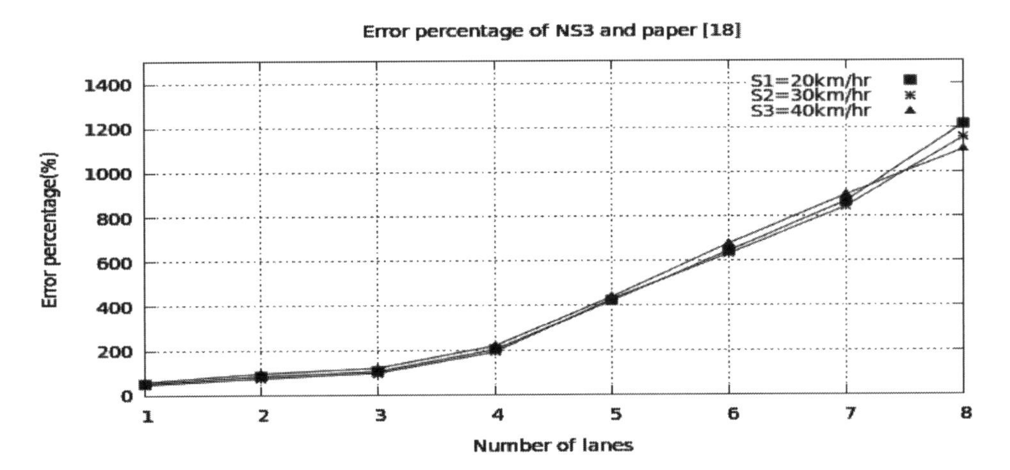

FIGURE 7.9
Error percentage in model of paper [17].

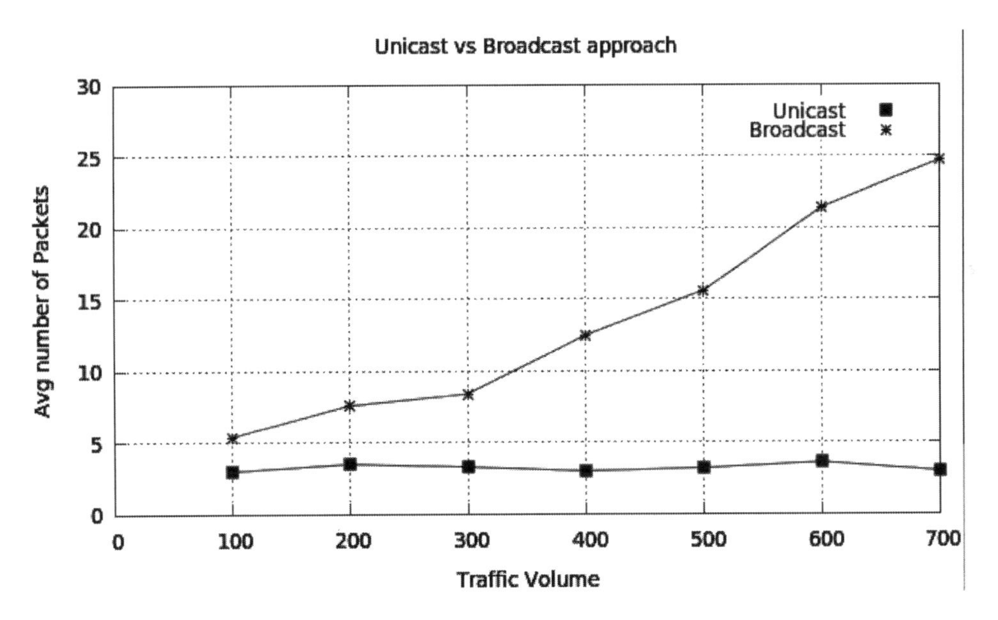

FIGURE 7.10
Unicast Vs broadcast routing approaches.

discovered procedure ranges from 0.5 to 3s, revealing the importance as well as the effectiveness of the system revealed in discovery of V2V neighbors in VANET network. When evaluating the established LTE-based network V2V discovery [13], the stated VANET network-based V2V revelation offers comparable delay range. In the event of LTE-based network, a V2V scheme offers two types of services: Type 1 and Type 2, each symbolizing varied attributes with regard to power consumption and delay. An enhancement in space between RSUs will contribute to an enhanced delay, though results examined have $W_{RSU} = 1200\ m$ and in range of RSU being 300 m are comparable to LTE-based development.

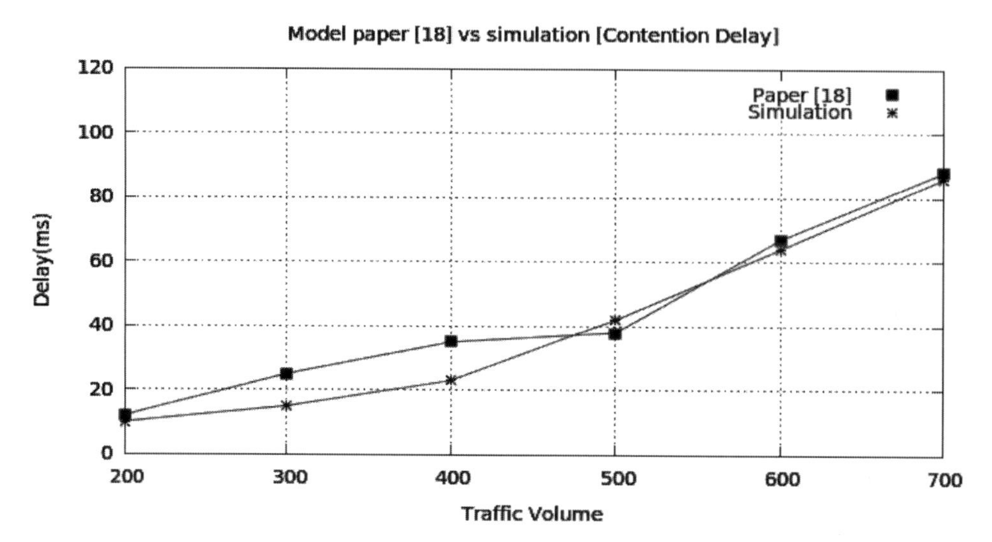

FIGURE 7.11
Model paper [17] Cs simulation (contention delay).

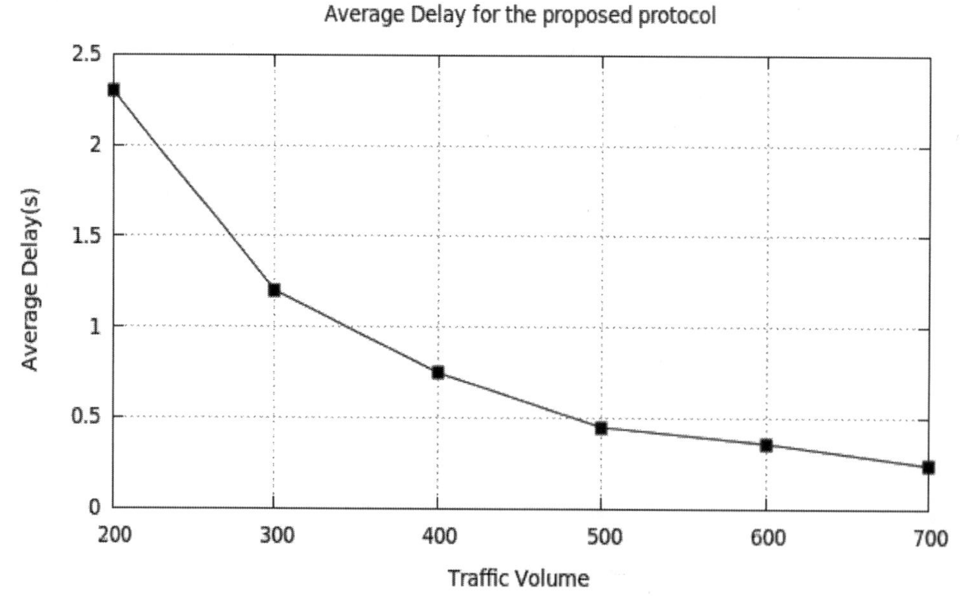

FIGURE 7.12
Average delay in explained protocol.

References

1. A. Vladyko, A. Khakimov, A. Muthanna, A. A. Ateya, and A. Koucheryavy, "Distributed edge computing to assist ultra-low-latency VANET applications," *Futur. Internet*, vol. 11, no. 6, 2019, doi: 10.3390/fi11060128

2. A. Arora, A. Mehra, and K. K. Mishra, "Vehicle to vehicle (V2V) VANET based analysis on waiting time and performance in LTE network," in Proceedings of the International Conference on Trends in Electronics and Informatics, ICOEI 2019, 2019, pp. 482–489, doi: 10.1109/ICOEI.2019.8862776
3. M. S. Sheikh and J. Liang, "A comprehensive survey on VANET security services in traffic management system," *Wirel. Commun. Mob. Comput.*, vol. 19, no. 5. 2019, doi: 10.1155/2019/2423915
4. J. Zhao, Z. Wu, Y. Wang, and X. Ma, "Adaptive optimization of QoS constraint transmission capacity of VANET," *Veh. Commun.*, vol. 17, no. 7, pp. 1–9, 2019, doi: 10.1016/j.vehcom.2019.03.005
5. Y. N. Liu, S. Z. Lv, M. Xie, Z. Bin Chen, and P. Wang, "Dynamic anonymous identity authentication (DAIA) scheme for VANET," *Int. J. Commun. Syst.*, vol. 32, no. 5, pp. 7–10, 2019, doi: 10.1002/dac.3892
6. R. Hussain, F. Hussain, and S. Zeadally, "Integration of VANET and 5G Security: A review of design and implementation issues," *Futur. Gener. Comput. Syst.*, vol. 101, no. 2, pp. 843–864, 2019, doi: 10.1016/j.future.2019.07.006
7. H. Hasrouny, A. E. Samhat, C. Bassil, and A. Laouiti, "Trust model for secure group leader-based communications in VANET," *Wirel. Networks*, vol. 25, no. 8, pp. 4639–4661, 2019, doi: 10.1007/s11276-018-1756-6
8. F. Goudarzi, H. Asgari, and H. S. Al-Raweshidy, "Traffic-aware VANET routing for city environments-a protocol based on ant colony optimization," *IEEE Syst. J.*, vol. 13, no. 1, pp. 571–581, 2019, doi: 10.1109/JSYST.2018.2806996
9. H. Zhong, S. Han, J. Cui, J. Zhang, and Y. Xu, "Privacy-preserving authentication scheme with full aggregation in VANET," *Inf. Sci. (Ny).*, vol. 476, no. 21, pp. 211–221, 2019, doi: 10.1016/j.ins.2018.10.021
10. Y. Xiao and Y. Liu, "BayesTrust and VehicleRank: Constructing an Implicit Web of Trust in VANET," *IEEE Trans. Veh. Technol.*, vol. 68, no. 3, pp. 2850–2864, 2019, doi: 10.1109/TVT.2019.2894056
11. H. Sedjelmaci, M. A. Messous, S. M. Senouci, and I. H. Brahmi, "Toward a lightweight and efficient UAV-aided VANET," *Trans. Emerg. Telecommun. Technol.*, vol. 30, no. 8, 2019, doi: 10.1002/ett.3520
12. S. Benkerdagh and C. Duvallet, "Cluster-based emergency message dissemination strategy for VANET using V2V communication," *Int. J. Commun. Syst.*, vol. 32, no. 5, 2019, doi: 10.1002/dac.3897
13. S. Kim, "Impacts of Mobility on Performance of Blockchain in VANET," *IEEE Access*, vol. 7, no. 2, pp. 68646–68655, 2019, doi: 10.1109/ACCESS.2019.2918411
14. A. Arora, N. Rakesh, and K. K. Mishra, "Asset distribution and peerless selection approach for LTE based multifarious VANET networks," *Int. J. Math. Eng. Manag. Sci.*, vol. 4, no. 1, pp. 27–40, 2019, doi: 10.33889/ijmems.2019.4.1-003
15. A. Arora, A. Mehra, K. K. Mishra, and N. Rakesh, "Strengthening connectivity in VANET by making use of driverless cars," *Int. J. Eng. Adv. Technol.*, vol. 8, no. 6, pp. 4890–4898, 2019, doi: 10.35940/ijeat.F9133.088619
16. A. Arora, A. Mehra, and K. K. Mishra, "Vehicle to vehicle (V2V) VANET based analysis on waiting time and performance in LTE network," in Proceedings of the International Conference on Trends in Electronics and Informatics, *ICOEI 2019*, 2019, pp. 482–489, doi: 10.1109/ICOEI.2019.8862776
17. M. B. Mansour, C. Salama, H. K. Mohamed, and S. A. Hammad, "VANET Security and Privacy - An Overview," *Int. J. Netw. Secur. Its Appl.*, vol. 10, no. 2, pp. 13–34, 2018, doi: 10.5121/ijnsa.2018.10202

8

Intelligent Routing in Vehicular Ad Hoc Networks

Rohit Srivastava

CONTENTS

8.1 Overview

A QoS-mindful versatile steering calculation (IGLAR, improved geolocation based QoS aware routing algorithm) is proposed in this work, which centers on an ideal way to ID successful directing over a profoundly dynamic portable impromptu network like VANET. Conventional directing conventions, for example, Greedy Perimeter Stateless Routing (GPSR), Ad-hoc On-demand Distance Vector (AODV) do not zero in on vehicle traffic conduct at any cross streets or intersections. Consequently, a steering convention to deal with the geospatial area of a vehicle and control weighty traffic power is required. The interaction of determination and using the ideal QoS course is refreshed on transmission.

An IGLAR is proposed between the sender and receiver, which is dependent on the QoS parameters and the interaction between the final intercepts depends on the communication between the two nodes.

The proposed plan can be defined as follows:

1. The major test in the convention plan in VANET is to improve the unwavering quality of directing information with help to decrease conveyance deferral time.

DOI: 10.1201/9781003102397-8

2. Abstract vehicular versatility is one of the significant issues that prompt delay, thus planning of postponement-limited steering conventions is likewise a test since multicast store and forward is a way to deal with conveying packets.
3. Priority for vehicles like ambulances and VIPs that require safe travel on exceptionally blocked street is considered in this examination.

To investigate the presentation of IGLAR, two diverse metropolitan street situations were considered. Continuous street routes were made with VANETMobiSim, which can chip away at the ns3 test system. The situation maps think about the fluctuating velocity of vehicles, vehicle need, administration need, and pass-through street intersections, path data, and traffic power on streets at shifting time stretches. The mimicked results show that the IGLAR convention's presentation is discovered to be superior to GPSR and AODV conventions for both situations. Execution of measurements like effective information conveyance and normal postponement were contemplated.

8.2 Theoretical Background

Remote correspondence is universal because of its adaptability to adjust to various situations. Mobile ad hoc networks (MANETS) is a term authored for the ceaselessly fluctuating network geography provided in handheld mobile gadgets. Vehicular ad hoc networks (VANETS) are one of those. It sends the idea of consistently differing vehicular movement. The hubs or vehicles in VANETS can move around with no limits on their course and speed. This discretionary movement of vehicles presents new difficulties to analysts as far as to find new route for all the more explicitly for VANETS. Gridlock on the streets today is an enormous issue in large urban areas. The blockage and related vehicle inconvenience is joined by a steady danger of mishaps. Absence of street traffic security incurs significant damage to human lives and represents a desperate danger. Other adverse results are identified with energy waste and ecological pollution.

8.2.1 VANET Architecture

As indicated in Figure 8.1, the architecture of VANETs falls into three classes: pure cell/WLAN, pure ad-hoc, and hybrid. In pure cell/WLAN architecture, the network

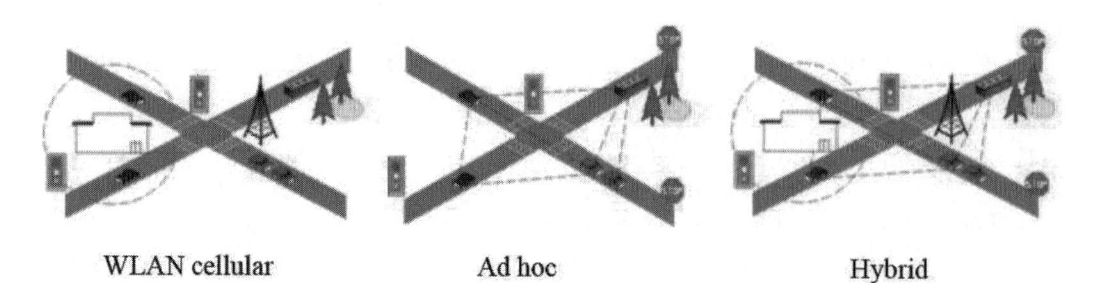

FIGURE 8.1
VANET architecture [5].

utilizes cell doors and WLAN passages to interface with the internet and work with vehicular applications. Vehicles speak with the internet by driving by either a cell tower or a remote passage. Since the foundation of the cell, pinnacles, and remote passages are not generally conveyed because of expense or geographic impediments, although hubs may correspond with one another. Data gathered from sensors on a vehicle can be significant in informing different vehicles about traffic conditions and assisting the police with tackling wrongdoings [4]. The framework less network architecture is in the pure impromptu class where hubs perform vehicle-to-vehicle correspondence with one another.

When there are side-of-the-road correspondence units, like a cell tower or a passageway, and vehicles are outfitted with remote system administration gadgets, vehicles can exploit the foundation in speaking with one another. Different applications in areas of metropolitan observation, security, driving help, and diversion have utilized foundation-conveying units to get dynamic and rich data outside their network setting and offer this data in a distributed design through impromptu, foundationless correspondence [4]. The half-breed architecture of cell/WLAN and impromptu methodologies gives more extravagant substance and more noteworthy adaptability in content sharing.

Like mobile ad hoc networks, hubs in VANETs also self-arrange and self-oversee data in a disseminated style without a concentrated power or a worker directing the correspondence. In this kind of network, hubs connect with themselves as workers as well as customers, trading and sharing data like companions. Besides, hubs are portable, consequently making information transmission less solid and problematic.

8.2.2 Characteristics of VANET

In planning conventions for the cutting-edge vehicular network, we perceive that hubs in these networks have essentially different qualities and requests from those customary in remote specially appointed networks conveyed in framework with fewer conditions (e.g., sensor field and combat zone). These distinctions have a critical sway on application foundations:

a. **High dynamic geography**
The speed and decision of route characterize the unique geography of VANET. On the off chance that we accept two vehicles moving away from one another with a speed of 60 mph (25 m/s) and if the transmission range is about 250 m, the connection between these two vehicles will keep going for just 5 seconds (250 m/50 ms^{-1}). This characterizes its profoundly powerful geography.

b. **Regular disengaged network**
The above highlight requires that in regular convention the hubs require another connection with a nearby vehicle to keep up consistent availability. Yet, if there should be an occurrence of a low vehicle thickness zone, incessant disturbance of network availability will happen. Such issues are on occasion tended to overcome the issues of overhead.

c. **Portability modeling and prediction**
The above highlights for the network in this way require information on hub positions and their developments, which is hard to anticipate. In any case, a portability model and hub expectation dependent on the investigation of a predefined streets model and vehicle speed are of principal significance for a powerful network plan.

d. **Correspondence environment**
The portability model exceptionally fluctuates from rural roads to city streets. The hub expectation plan and steering calculation in this way need to adjust for these changes. The thruway versatility model, which is a one-dimensional model, is basic and simple to anticipate. However, for the city portability model, road structure, variable hub thickness, presence of structures, and trees that act as impediments to even little distance correspondence make the model application exceptionally perplexing and troublesome.

e. **Connection with installed sensors**
This sensor helps in giving hub areas and their development a nature that is utilized for successful correspondence connection and directing purposes.

8.3 Literature Survey

The work is evaluated in independent segments giving subtleties of various steering conventions and their correlation. In VANET, there are numerous applications that use directing conventions like map identification for drivers, traffic security, mishap aversion, traffic information dispersal, and so forth. More emphasis is given to directing conventions and their correlation as it is the point of the present investigation.

Jagadeesh Kakarla, S. Siva Sathya, B. Govinda Laxmi, and Ramesh Babu present "A Review on Routing Protocols and its Issues in VANET." This paper gives a short outline of various directing calculations in VANET alongside significant orders. The conventions are likewise thought about dependent on their fundamental attributes and network. Vehicular specially appointed networks offer countless applications without any help from a fixed framework.

VANET applications are exceptionally hard. Henceforth an overview of various VANET conventions, contrasting the different highlights is significant to concoct a new proposition for VANET. The presentation of VANET steering conventions relies upon different boundaries like portability model, driving climate, and some more. Along these lines, this paper has come up with a thorough study and correlation of various classes of VANET steering conventions. From the study, unmistakably position-based group-based conventions are more dependable for the greater part of the applications in VANET.

Monika Khatri and Sona Malhotra address a paper on "Conduct investigation of Steering Protocols in VANET." In this paper, the creator endeavors to distinguish significant issues and difficulties related to various VANET conventions and to choose the ideal VANET convention for the forecast of the future.

A few startlingly grievous circumstances are experienced on street networks every day, a significant number of which may prompt blockage and well-being dangers. On the off chance that vehicles can be given data about such episodes or traffic conditions ahead of time, the nature of driving can be improved as far as time, distance, and well-being. One of the fundamental difficulties in a vehicular ad hoc network is of finding and keeping a viable course for shipping information data. The creator will endeavor to discover significant issues and challenges related to various VANET conventions, and to choose the ideal VANET convention for the expectation of the future. The end of this paper is that it manages the investigation of various types of VANET steering conventions. It decides the application and attributes of vehicular ad hoc networks. After that, a point-by-point investigation of steering conventions and their correlation in various insights is analyzed. Generally, the overview sums up that steering convention of VANET needs improvement in steering traffic load, throughput, start-to-finish delays, control overhead, and handoff and meeting network time. This paper features issues and difficulties that might be useful for the future specialists to carry on their work.

Gayathri Chandrasekaran, in her paper "The Networking Platform for Future Vehicular Applications," takes the position that VANET would turn out to be the systems administration stage that would uphold the future vehicular applications. She dissects the components that are basic in choosing the systems administration structure over which the future vehicular applications would be sent and shows that there are dynamic research endeavors towards making VANETs a reality soon. She centers on well-being and internet connectivity related applications in this paper. At the end of this paper, the author contended that VANETs would end up being the systems administration foundation for supporting future vehicular applications. She began with depicting the variables that would be basic in making VANETs a reality, followed by a conversation on the exploration challenges. She showed that there are a few tests, including security and protection, and that dynamic exploration endeavors are being attempted to connect the holes needed to make VANETs a reality. Lastly, she talked about the counter cases that tested the reasonableness of VANETs and showed that there are solid purposes behind vehicular applications to be sent and that pure V2V- or V2I-based arrangements won't be adequate and VANETs will undoubtedly prevail in obliging these applications.

Valery Naumov and R. Thomas wrote "Net proposed "Network-Aware Routing (Vehicle) in Vehicular Ad Hoc Networks" that shows that position-based directing conventions such as Dynamic MANET on demand (DYMO), GPSR need earlier information on geographic area data of vehicles (from a GPS administration). This could be applied in VANETs for quicker course data and execution. It had been seen that position-based directing conventions likewise have serious geographic directing disappointments because of the quality of "geography openings," The authors propose spatially mindful steering to beat such disadvantages. In any case, the viability in the optimality of spatially mindful directing could not be decided because of spatial non-mindfulness. Thus it very well may be demonstrated and it very well may be additionally upgraded to improve execution. It follows a pestilence steering approach for VANET with resistance support for the delay. In inadequate vehicular rush hour gridlock, intelligent sending systems would be useful for vehicle-to-vehicle specially appointed interchanges. Aqua Sim (AQVA) steering convention is dynamic in distinguishing the final boundaries. This plan had been reproduced utilizing Vanet-Mobisim over ns3, with help for the constant street situation. The major disadvantage of this AQVA is that it is deficient in keeping up course which is one of the significant necessities of media web-based features on the side of QoS.

8.4 Proposed Work

IGLAR plans and reproductions are principally founded on the vehicular conduct and correspondence model. The ns-2 reenactments at first completed over IEEE 802.11 for VANET demonstrate the outcome of the overhead for the representation of protocol in which the final outcome of the route depends on the rote estimation (Fig. 8.2).

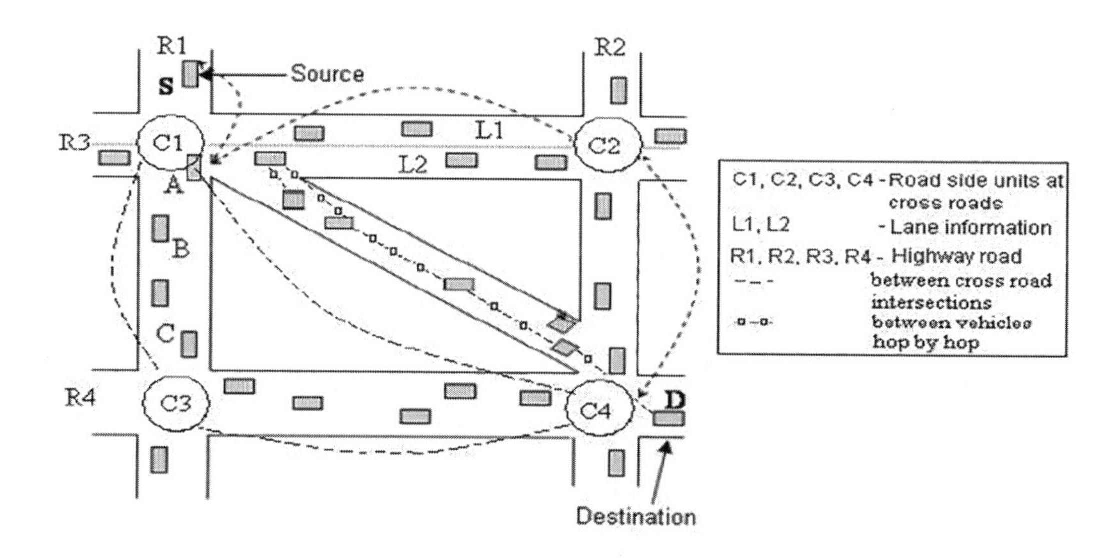

FIGURE 8.2
IGLAR design [2].

8.4.1 Infrastructure Model

IGLAR, as in Figure 8.2 is displayed as a bunch of high-velocity vehicles on a straight roadway in which any vehicle can develop availability with some other vehicle(s) going in the same heading or inverse bearing of its movement.

The traffic power of vehicles at any moment – t, where t can be accumulated at any Ci or Ri or Li values of the route. The power of vehicles on the street or between lane(s) decides the decency of giving the QoS on request. As IGLAR receives DSRC standard, IEEE 802.11 MAC is considered at interface layer imparting at 5.9 GHz recurrence. The RSU hubs at cross streets control the power of vehicular traffic over intermittent periods. An association set up between the sources „S and objective or beneficiary „D can be given as bounce by-jump association utilizing sending hubs „Fi or utilizing hubs at cross streets „Ci. Any „Ci can keep up the personality of vehicles inside its radio correspondence range and can speak with its neighbor „Cj or „Ck. Vehicles can be delegated source hub „S or objective hub „D, and can advance information during travel as sending hub „Fi.

8.5 Implementation Methodology

Three significant functionalities are completed by a conventional ad hoc directing convention, such as the revelation of a new course, choosing an ideal course (from numerous legitimate courses accessible), performing course support for the movement of information during travel, and providing course updates. IGLAR additionally receives the accompanying course that are required for QoS.

IGLAR deals with course-creation measures, where a course is thought of as ideal only if it can fulfill the necessary IGLAR QoS measurements. On the off chance that an ideal

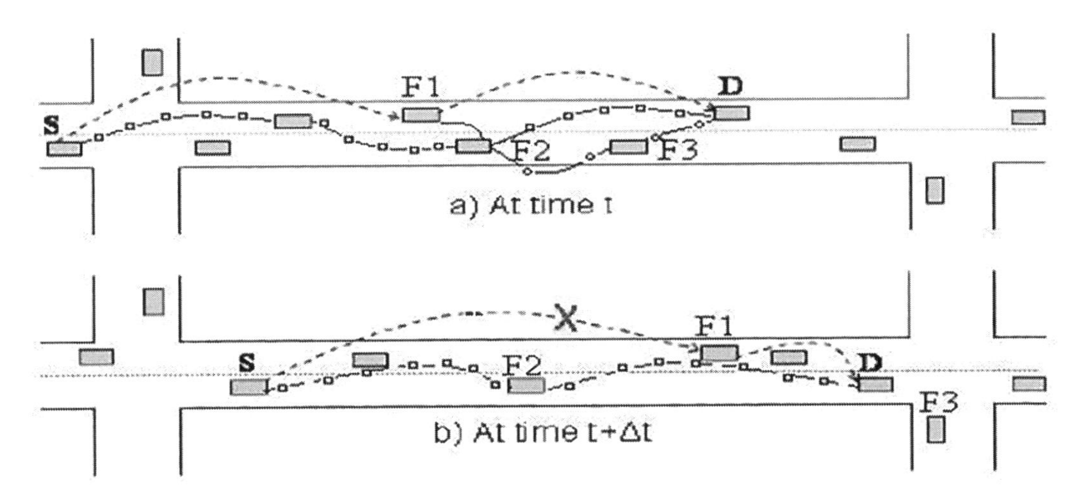

FIGURE 8.3
Vehicle Communication at different time intervals [4].

course has been recognized from a source to its objective, course support strategy ought to be completed to screen the meeting being used.

QoS parameter is referred for many issues like delay, execution, etc. in IGLAR for the parameters that are referred in the route tracing between sender and receiver (Fig. 8.3). Course upkeep technique and re-disclosure measures include broad flagging and calculation strategies. Consequently, the attractive alternative is to choose the ideal course with connections of the greatest potential lifetime during the ideal course choice stage.

Course optimality needs a higher need alongside interface delay since it is more crucial for the course quality in vehicular conditions.

a. At time „t, Route [S,F1,D] and backup course of action [S,F2,F3,D].
b. At time „t, Route [S,F2,D], which is an optimal course because of disappointment of course [S,F1,D].

8.5.1 Creating Routes

Procedure 1 {Route Create (Route_ID, Route Next, QoS_value)}

Route Request (IGLAR_REQ) and Route Reply (IGLAR_RPL) at any node Fi

Variables:

S, D: Identity of source and destination VANET nodes

Route [ ]: Array route consisting of all temporary VANET nodes

Route_OPT, TempRoute: Optimal route and temporary routes from S to D

ζ: Vehicular priority

|Hopk|: 'k' number of hops between S to D, where 'k' being the radio propagation length

Ri (Li, Fi): Road segment with Lane segment where VANET node Fi is located

```
Ci = Cross-road route
τ: Route update Time Wait (TW) parameter
IGLAR_REQ: Route request packet
IGLAR_RPL: Route reply packet
IGLAR_OPT: Optimal Route
Upon receiving IGLAR_REQ (S, D, TempRoute) from any Fi
1: if (S == D) & (|TempRoute| ∊ Route) then
2: Route OPT = TempRoute
3: Send IGLAR_RPL(S, D, Route_OPT)
4: return
5: else
6: send IGLAR_RER (0)
7: end if
8: if IGLAR_REQ = θ
9: if (Ri (Fi) ≠ Ri (Fj)&(Ri(Fi) ⊂ TempRoute) then
10: add Ri (Fi) to Route []
11: end if
12: set Hop k = distance (Fi, Fj) τ
13: increment Hop k
14: endif
15: if Ri(S) == Rj(D) then
16: stop Hopk /* Fi is a better broadcast node*/
17: end if
18: set τ = 0
19: IGLAR_BCS Route (S, D, TempRoute) /* broadcast route */
20: receive IGLAR_RPL (D, S, Route_RPL (Fj-1, Fi-1, -1)) from Fj
21: if τ ≠ -1 then goto step 8
22: else
23: continue
24: if (Fi == S) then
25: store Route_RPL in Ci
26: forward IGLAR_REQ (S, Fi, ROUTE_RPL(Fi+1, Fj+1, D,ζ))
27: end if
```

8.5.2 Route Discovery and Update

Identifying an optimal route for service between a source and a destination defines the process of satisfying the QoS on demand as per IGLAR metrics. Any service can be effectively accomplished if a best possible route or an optimal route among the available links is selected. The "capability" of defining an optimal route is based on the communication effectiveness for expected service in terms of fuzzy measure. Any node or link that is not "capable" to communicate as per optimality condition is defined as worst.

Optimization helps in providing an adaptive service for services that demand QoS consistently such as streaming media delivery, content management feed, and media conference. Optimization is provided for (a) assigning a route with required bandwidth, (b) maintaining and monitoring IGLAR metric ton delay, packet loss, number of helps, and radio propagation range.

Procedure 2: Optimal Route

IGLAR_OPR(NodeID_send(j),NodeID_recv(j), IGLAR_metric, Link_ID(j), ζ, μ,), where j is set of route links identified between 1 to n

Variables:

S, D: Identity of source and destination VANET nodes

Route [ ]: Array route consisting of all temporary VANET nodes

Rote_OPT, TempRoute: Optimal route and temporary routes from S to D

ζ: Vehicular priority

μ: Service priority

|Hopk|: 'k' number of hops between S to D, where 'k' being the radio propagation length

Ri (Li, Fi): Road segment with Lane segment where VANET node Fi is located

Ci: = Cross-road route

τ: Route update Time Wait (TW) parameter

IGLAR_REQ: Route request packet

IGLAR_RPL: Route reply packet

IGLAR_OPT: Optimal Route

Upon receiving IGLAR_REQ (S, D, TempRoute) from any Fi

1: if ((S == Fi) || (D==Fj) & (|TempRoute| ∊ Route [ ]) then

2: Route_OPT = TempRoute

3: send IGLAR_RPL(S, D, Route_OPT)

4: return

5: else

6: send IGLAR_REQ(S, D, TempRoute, μ, ζ)

7: set Hopk = distance(Fi, Fj) τ /* hop count between nodes */

8: set ζ = High ||Low|| Normal

9: set μ = High ||Low|| Normal

10: set τ = 0

11: if IGLAR_OPT = θ

12: if (Ri (Fi) ≠ Ri (Fj)&(Ri(Fi) ⊂ TempRoute) & IGLAR_REQ (Fi+1,Fj+1, μ) then

13: add Ri (Fi) to Route_OPT /* add the best route to Optimal Route */

14: end if

```
15: increment Hopk
16: if RI (S) == Rj (D) then
17: stop Hopk                       /* Fi is a better broadcast node */
18: end if
19: send IGLAR_OPT (S, D, TempRout,ζ) /* Optimal route */
20: receive IGLAR_RPL (D, S, Route_RPL (Fj-1, Fi-1, -1)) from Fj
21: increment IGLAR_OPT
22: if τ > 1 then goto step 10
23: else
24: continue
25: endif
26: if (Fi == S) then
27: store IGLAR_OPT in Ci and Ri
28: forward IGLAR_REQ (S, Fi, ROUTE_RPL (Fi+1, Fj+1, D,ζ))
29: end if
30: endif
```

The step-by-step explanation of the algorithm is discussed in this section. Steps 1 to 6 explain the optimal route identified if the route is found to be shortest between the source and destination, with no other possible routes found in *TempRoute* list. Steps 7 to 10 assign default values for IGLAR_OPT metrics. Step 11 checks whether an optimal route is available in list IGLAR_OPT, or the process of adding the possible links based on the service request is added to IGLAR_OPT as explained in Step 12 to 14.

8.5.3 Message Flow

The wonders of recognizing the ideal course IGLAR_OPT are profoundly subject to "Hopk" – the number of jumps chosen between sources and objective "D," μ and z, the vehicle, and the administration needs separately. On request, the transfer speed can be furnished to hubs and administrations with high need with the assistance of VANET hubs, which is talked about in steps 15 to 18. Stage 19 demands the ideal course for meeting foundation, for which answer is obtained and the meeting is set up. If the stand-by period t is extreme, the break is pronounced, and the meeting is reset as demonstrated in Steps 22 to 25. The best ideal course is put in cross street hub Ci or Ri (Step 27), while demand is sent for a meeting course update in Step 28. The stream graph of IGLAR messages dealing with schedules is demonstrated in Figure 8.4.

This interaction is done with IGLAR_REQ and IGLAR_RER for connection establishment. IGLAR_RER is produced when a hub or connection had fizzled during transmission; the bundles sent to the bombed hub must be educated to the neighbor sending hub Ci. The sending hub Ci at cross streets re-communicates or communicates the information to keep away from irregular deferral at the recipient end.

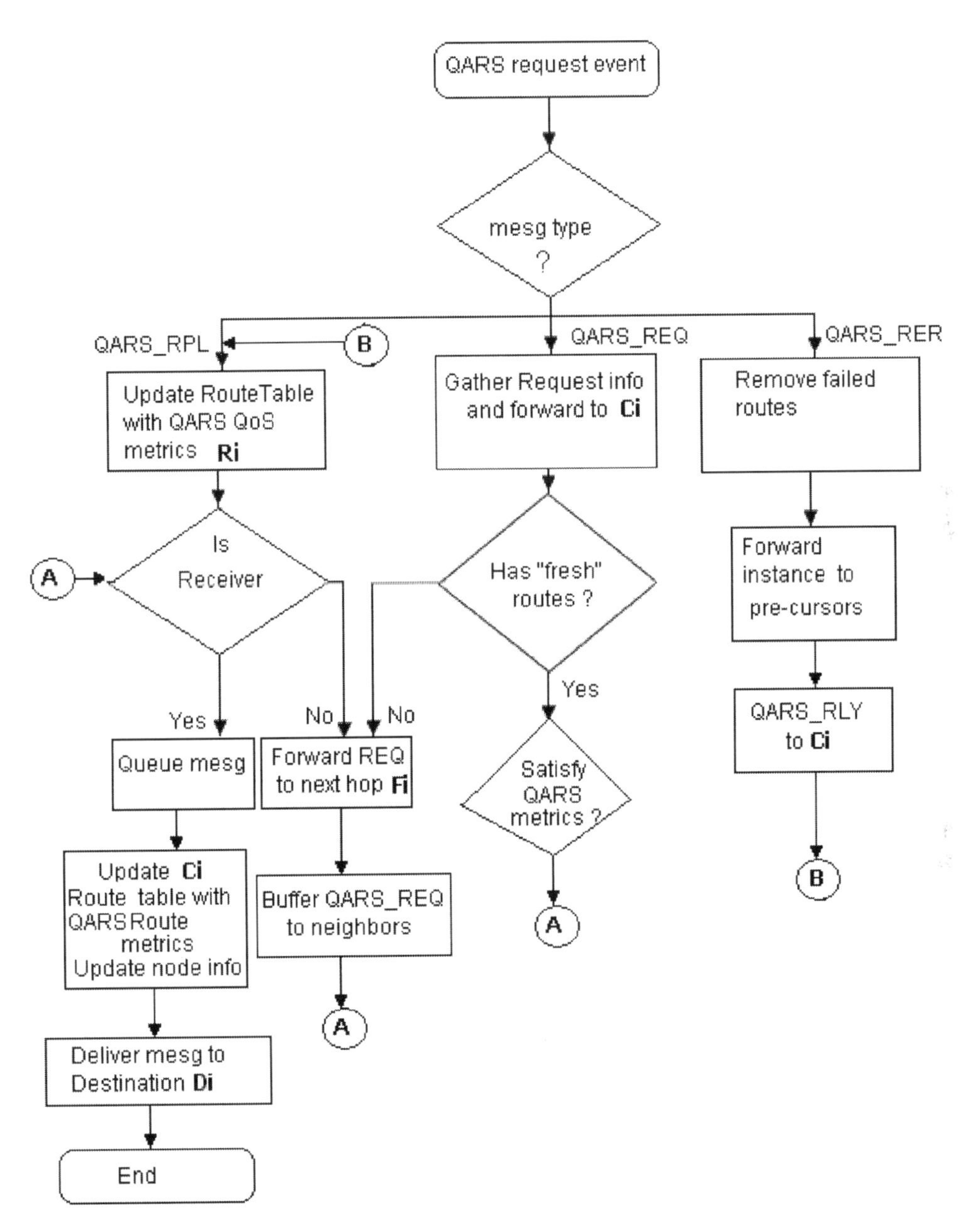

FIGURE 8.4
Message flow in IGLAR.

8.6 Performance Analysis

The recreated conduct of vehicles can be imagined in Fig. 8.4 with explicit QoS boundaries, for example, "wanted speed," "course of the vehicle," "change of path data," which were utilized to display various sorts of street clients. VANET hubs were characterized into the type of truck that can go at the most extreme speed of 22.2 m/s (approx. 60 to 80 km/h) and car that can go at a normal speed of 33.0 m/s (approx. 80 to 120 km/hr.).

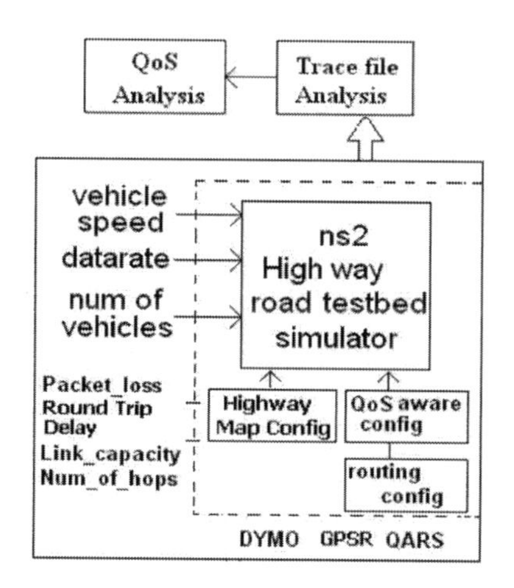

FIGURE 8.5
Simulation setup [6].

To show the proving ground situation, two diverse contextual analyses were considered. The execution of the IGLAR convention is assessed in high-dispute conditions (Fig. 8.5):

1. A highway metropolitan street climate with significant hindrances, which message utilizing occasional "hi" and the standard 802.11 MAC convention. VANETMobisim upholds the street map against genuine vehicular follows along the street.
2. A metropolitan climate without deterrents, utilizing IGLAR's proposed sending enhancement steering plan, across the streets.

A vehicular traffic generator "vehicle following model" proposed by Gipps, had been proposed. This model empowers vehicles "on versatility" with the end goal that it can move from the most extreme, most secure speed to the lowest speed to dodge crashes. However, the impediment of this model is that it doesn't uphold constant situation guides; thus as another substitute, VANETMobisim is utilized for traffic calculation (Fig. 8.6). The proving ground is

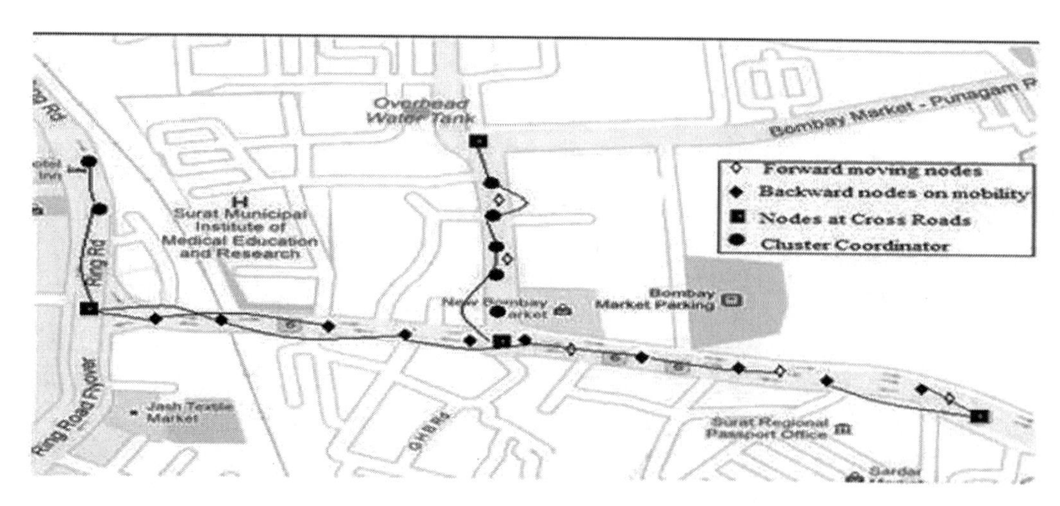

FIGURE 8.6
Surat city digital road map [5].

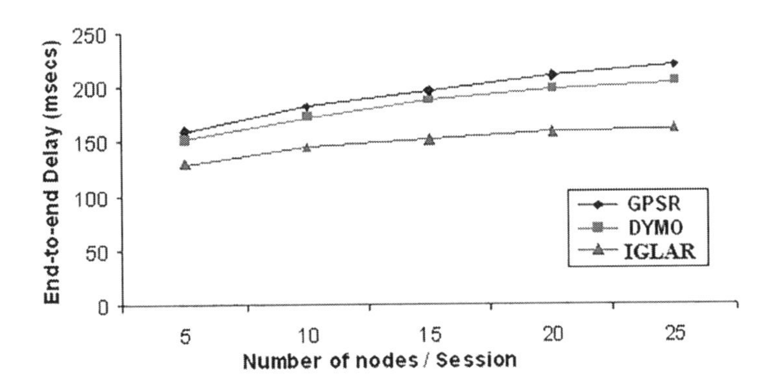

FIGURE 8.7
End-to-end delay.

executed utilizing a unique situation; the information rate is booked with changing vehicular speed utilized on interstate streets. The boundaries characterized are according to the standard received in IEEE 802.11 Macintosh support for several hubs and transmission/receiver range.

The execution was estimated at different VANET hub densities, for example, of 4, 7, 15, and 25 vehicles for each kilometer and path, comparing to selected traffic densities along with a highway guide. Vehicles are ordered into "high priority," "rescue vehicle," and "ordinary priority." The fig likewise shows cross-street hubs, which convey shareable data transfer capacity, which hubs can share for QoS on request benefits. Ambulances and high-need hubs were given higher IGLAR QoS support contrasting with different hubs, which used the cross-street hub more for using a lot of transfer speed. The normal priority hubs followed behind and utilized the middle-of-the-road forward hubs for transmission.

Figure 8.7 shows the presentation of IGLAR in examination with the reviewed convention. It was seen that delay for GPSR and DYMO was normally 12%, in high contrast with IGLAR, as the number of hubs was gradually expanded to 25. The presentation of IGLAR varies undeniably, with the delay of about 90 ms. Normally IGLAR performs with a start-to-finish deferral of 125 ms, while DYMO shows 192 ms and GPSR 210 ms.

Figure 8.8 shows the presentation of IGLAR as PDR (parcel conveyance proportion), which clarifies the throughput or bundle misfortune proportion (in %). From this figure, it tends to be perceived that the exhibition of IGLAR and DYMO was high at first (normal

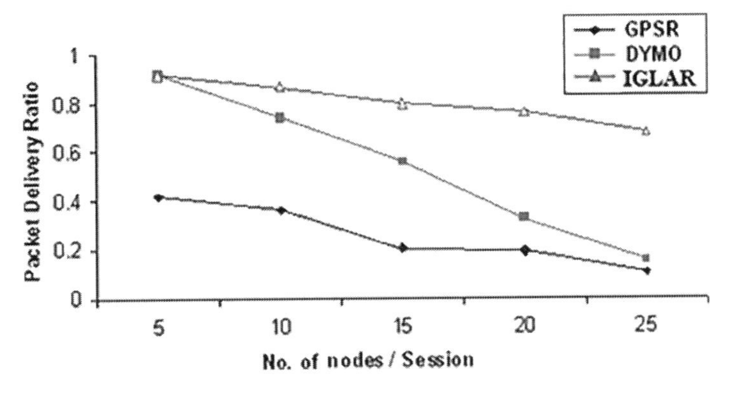

FIGURE 8.8
Packet delivery ratio.

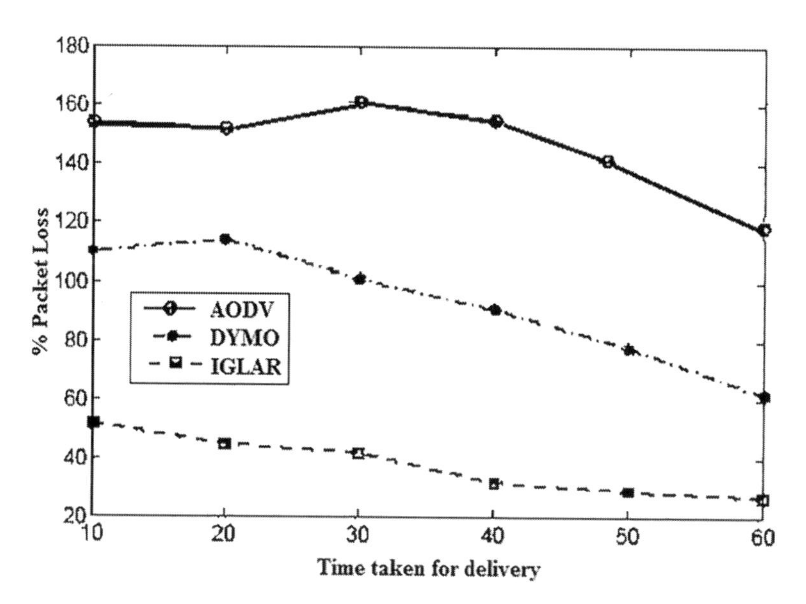

FIGURE 8.9
Packet loss at cross nodes.

100%), while GPSR conveyed just 40% of sent information. The number of hubs was expanded, it was discovered that IGLAR shows a deficiency of 20% of its information, yet DYMO and GPSR lose around 90% of the information. The exhibition of IGLAR is slightly better compared to DYMO and GPSR.

Figure 8.9 exhibits the level of bundle misfortune accomplished at transitional cross streets where the presentation of IGLAR was better compared to contrast with DYMO and AODV.

8.7 Conclusion and Future Work

This exploration work of IGLAR centers around giving QoS on request to vehicles that chip away at data transfer capacity hungry applications. IGLAR directing conventions deal with the upkeep of a start-to-finish way for streaming media information to arrive at its objective. Significant commitment from this examination work centers on giving QoS to "on request benefits" for a vehicle on powerful versatility. Efforts on controlling vehicles with variable speed and path change and vehicles at cross streets are grounded through recreated results. So far, vehicle communication and administration are not examined; henceforth such works are not taken as a component for study.

Future work can be towards improving the unwavering quality and noteworthiness, which is the live test in the convention plan in VANET. A continuous investigation on VANETs and trial approach ought to be received to improve quality. Viewpoints on driver conduct (sleepiness, drunk driving, and thoughtlessness) ought to be considered for planning of several conventions for handling overhead. IGLAR can be improved in the future to help with continuous web-based features.

References

1. B. Karp and H. T. Kung, "GPSR: Greedy perimeter stateless routing for wireless networks,", in Proceedings of the ACM/IEEE International Conference on Mobile Computing and Networking (MobiCom), 2000.
2. C. Lochert, M. Mauve, H. Fubler, and H. Hartenstein, "Geographic routing in city scenarios," ACM SIGMOBILE Mobile Computing and Communications Review (MC2R) [C], vol. 9, no. 1, pp. 69–72, Jan 2005.
3. D. Johnson, B. D. A. Maltz, and Y. C. Hu, "The Dynamic Source Routing Protocol for Mobile Ad Hoc Networks (DSR)," draft-ietf-manet-dsr-10.txt, 2004.
4. F. Ros, P. M. Ruiz, and I. Stojmenovic, "Reliable and efficient broadcasting in vehicular ad hoc networks,, IEEE the 69th Vehicular Technology Conference (VTC'09), Spain, April, 2009.
5. F. Li and Y. Wang, "Routing in Vehicular Ad Hoc Networks: A Survey," IEEE Vehicular Technology Magazine, vol. 2, no. 2, pp. 12–22, June 2007.
6. X. H. Gerla and G. Pei, "Fisheye State Routing Protocol (FSR)," IETF Internet Draft, work in progress, draftietfmanet-fsr-03.txt, July 2002.
7. M. Gerlach, (2006). Full paper: assessing and improving privacy in VANETs.
8. H. Rahbar, K. S. Naik, and A. Nayak, "DTSG: Dynamic Time-Stable Geocast Routing in Vehicular Ad Hoc Networks," IEEE Symposium on Computers and Communications, pp. 198–203, 2001.
9. C. Harsch, A. Festag, and P. Papadimitratos, (2007). Secure position-based routing for VANETs. In Proceedings of IEEE 66th vehicular technology conference (VTC-2007), Fall 2007 (pp. 26–30), September 2007.
10. M. Jerbi, R. Meraihi, S.-M. Senouci, and Y. Ghamri-Doudane ENSIIE, "GyTAR: Improved Greedy Traffic Aware Routing Protocol for Vehicular Ad Hoc Networks in City Environments," VANET'06, September 2006.
11. C. Perkins, E. Belding-Royer, and S. Das, "Ad Hoc On-Demand Distance Vector (AODV) Routing," RFC 3561, Network Working Group, 2003.
12. R. A. Santns, R. M. Edwards, A. Edwards and D. Belis, "A novel cluster-based location routing algorithm for inter vehicular communication," Personal, Indoor and Mobile Radio Communications, IEEE proceedings of the 15th Annual Symposium, 2004.
13. M. Raya, and J. Hubaux, (2005). The security of vehicular ad hoc networks. In Proceedings of the 3rd ACM workshop on security of ad hoc and sensor networks (SASN 2005) (pp. 1–11)
14. Kevin C. Lee, Uichin Lee, Mario Gerla, Survey of Routing Protocols in Vehicular Ad Hoc Neworks, Advances in Vehicular Ad-Hoc Networks: Developments and Challenges, IGI Global, Oct, 2009.
15. T. Clausen, et al., "Optimized Link State Routing Protocol (OLSR)," RFC 3626, Network Working Group, Oct. 2003.
16. V. Park, S. Corson, "Temporally-Ordered Routing Algorithm (TORA) Version 1 Functional Specification," IETF Internet draft, work in progress, draft-ietf-manettoraspec-04.txt, 2001.
17. V. Naumov and R. Thomas, "Gross proposed "Connectivity-Aware Routing (CAR) in Vehicular Ad Hoc Networks"
18. Y. Luo, W. Zhang, Y. Hu, "A New Cluster Based Routing Protocol for VANET," IEEE Wireless Communications and Trusted Computing, 2010.
19. Z. J. Haas, "The Zone Routing Protocol (ZRP) for ad hoc networks," Internet Draft, Nov. 1997.

9

WSN Security Intrusion Detection Approaches Using Machine Learning

Gaytri Bakshi and Himanshu Sahu

CONTENTS

DOI: 10.1201/9781003102397-9

9.1 Introduction

A sensor (Raj et al. 2012) is the fundamental entity, which helps humans to interact and digitally understand the physical world by tracking the changes occurring due to arising events in certain specified areas. To study the phenomenal changing aspects of the environment as per the application, a collection of sensor nodes is used. These sensor nodes together form a wireless sensor network. WSN is actually a remote network in which a lot of aspects exists when it is deployed; one of these is security.

Security has become an important issue for any organization (either government or private) due to excessive use of internet and communication technologies (ICT). The emergence of ICT has increased the pace of socioeconomic development as well as growth of organizations, but it has also opened the door for the adversary. The problem is growing day by day with the emergence of cloud computing, mobile computing, and machine learning since the data is leaving the organization for processing. For any organization (and its constituent system) to detect, prevent, and mitigate malicious attacks, it is necessary to detect the cause, amend it, and make the entire system robust and reliable enough that it could resist an attack and have a self-correcting mechanism to improve the status and condition of the system. Although not every solution is perfect and tamper proof, it should be reactive and updatable as per new threats.

The hampering of devices or systems starts with the intrusion through the vulnerabilities associated with them. The vulnerabilities are the loopholes in the system design, such as insecure remote access and unused ports. These vulnerabilities could be categorized mainly into two aspects: the physical scenario and the digital world.

9.1.1 CIA Triad

To provide the information security within an organization the CIA triads must be preserved. CIA stands for confidentiality, integrity and availability. The organization's cyber-physical system and its security solutions should be designed to ensure the protection of the CIA triad. The CIA triad with its parts are shown in Figure 9.1.

9.1.2 Intrusion Detection System (IDS)

The intrusion detection system works on pattern analysis of network and resource usage. It may be based on the combination of hardware and software. An intrusion detection system has a very prominent role in providing cyber security to an organization. Modern intrusion detection systems employ machine learning, deep learning, and data analytics techniques.

9.2 Security and Intrusion

For any organization, cyber security has become an important infrastructural component due to excessive use of ICT and automation. Any organization – big or small, government or private – is dependent on ICT for running its day-to-day activities. ICT includes cloud computing, mobile computing and IoT, which has become part of every organization. Apart from ICT, artificial intelligence (AI) also plays a significant role to provide automation. AI techniques such as machine learning (ML) and deep learning (DL) are highly data dependent. An increase in cyber-physical attack leads to development of strong cyber security mechanisms to prevent and recover from such attacks. Any organization needs to maintain security the confidentiality, integrity, and availability.

9.2.1 CIA Triads (Stallings 2006)

CIA stands for confidentiality, integrity, and availability which comprises the security requirement of an organization. These three properties are the heart of security of any organization or computer.

9.2.1.1 Confidentiality

Confidentially stands for the protection of sensitive information and data related to an organization, its employees, or clients. Confidentiality covers two aspects: one is data protection, and the other is data privacy. Data protection refers to the prevention of unauthorized access to the user's private data and ensures access by the intended user only. Data privacy is related to the storage and usage of the information of a user by the organization or a third-party service provider, such as cloud data centers. DL and ML techniques use cloud services to process the data, which creates a data privacy risk by taking the organization's sensitive data outside the organization.

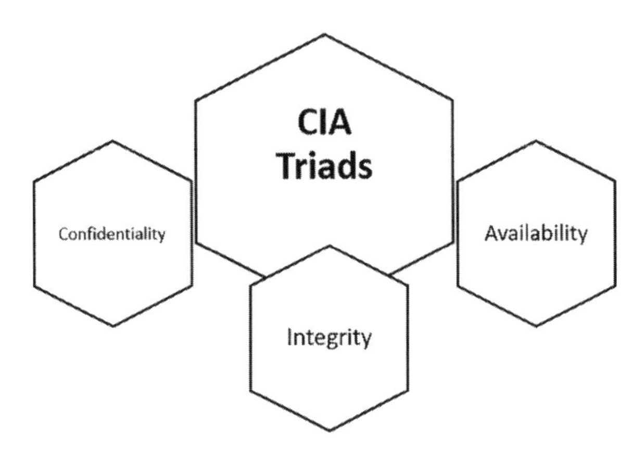

FIGURE 9.1
CIA triads.

9.2.1.2 Integrity

Integrity is related to protection from unauthorized changes in the user's data. Integrity is required to maintain data as well as the system. Data integrity refers to protection against unauthorized changes of user's data and programs. System integrity refers to protection against system malfunction (i.e., the system should be protected from change so that it can perform its intended task).

9.2.1.3 Availability

Availability refers to a system's ability to provide continuous services to the client up to its maximum capacity. A web server and file server should always provide services as defined in the service level agreement (SLA). High availability is the same as the uptime of a system.

9.2.2 Intrusion

National Institute of Standards & Technology describes intrusion as an attempt to compromise CIA or to bypass the security mechanisms of a computer or network. Intrusion detection is the process of monitoring the events occurring in a computer system or network and analyzing them for signs of intrusions (Liao et al. 2013). The origin of intrusion can be inside or outside.

In an insider attack the attacker is within the premises, such as when a dissatisfied or malicious employee or attacker successfully breaches physical security. In an outsider attack a special attempt to breach the network security to steal information or destroy the reputation is done remotely, such as by a cybercriminal, hacker, cracker, or hacktivist (Crume 2000).

9.2.2.1 Vulnerability

Vulnerability refers to a system's existing weakness caused by physical, environmental, economic, and social factors. Due to these causes, an intruder can gain access to any system – whether it is a cyber or physical platform – and perform spiteful tasks.

Various types of vulnerabilities are as follows:

a. Physical Vulnerability: This type of vulnerability includes the site, situation, configuration, design, and material required for the development of the respective physical or cyber system. Systems with such deficiencies are open to vandalism attacks. The classes of assets that can be encompass under this are as follows:
    - Weakness related to hardware includes the lack of security for unfortified storage.
    - Defenselessness related to software includes design imperfections and insufficient testing and auditing.
    - Exposure is related to organizational assets that are subject to unreliable power source and floods.
    - Vulnerability related to network includes unguarded communication lines and precarious network architecture.
b. Social vulnerability: This sort of vulnerability is because of lack of awareness regarding correct usage of social sites by highly active customers. This may lead

to disastrous and nasty activities. Attacks such as social engineering, honey traps, phishing, pretexting, baiting, quid pro quo, and tailgating target innocent people and demolish the entire system. The classes of assets that can be covered under this are as follows:

- Weakening the system due to lack of protocols in the recruiting process, which eventually leads to hiring untrained and unprofessional staff. This ultimately leads to wastage and wrong usage of network assets, which in due course heads towards a trap and distorts the entire system.
- Vulnerability within a system even occurs due to discontinuation and disruption in training the staff. This will lead to incompetency to tackle any cyber attack.

c. Economic vulnerability: This is one of the major impactful exposures, which heads towards a disadvantage of the organization as finance plays a major role in structuring the system to withstand the adverse impacts of malicious people and their activities.

The classes of assets that can be covered under this are as follows:

- Vulnerability related to lack of regular organizational audits may lead to inconsistencies of security resources, which may lead to downfall of the network of the organization.
- Vulnerability related to lack of continuous plan may head towards inconsistency in proper functioning of the system, which may lead to unprotected and unguarded resources, which further could be a weak link for the system to crash.
- Vulnerability related to lack of security in both the physical and cyber aspects can invite big attack to destroy the system.

d. Environmental vulnerability: This too is one of the key weaknesses that exist within the system as it is caused by nature and cannot be controlled; but it can be prevented to some extent.

9.2.2.2 Intruders

An intruder is an individual that attempts or gains unauthorized access to a cyber-physical system to exploit the system resource and data with malicious intent. The intruder can be a hacker or cracker. The hacker and cracker both work on vulnerabilities of the system but with subtle difference. The motive of the hacker is to find the vulnerabilities of the system, whereas the cracker tries to exploit the vulnerabilities of the system.

The intruders are categorized in three types:

1. Masquerader: An unauthorized individual who breaks into the system or network and exploits the system resources by gaining access to the account of the legitimate user.
2. Misfeasor: An individual with authorization to access the system and its resource but whose attempts go beyond their access limit.
3. Clandestine user: A user who has super user or administrative access and evades the monitoring. Activities done by such a user usually remain unidentified due to the administrative access by which they suppress the security and monitoring system

9.2.2.3 Intrusion Detection

Intrusion detection is a class of problems related to attack detection over an organization's system or an individual system. Intrusion detection is an attempt to identify whether the system is working normally. It works by finding abnormalities in system usage or user's behavior. Deviation from normal behavior can lead to attempts to exploit the vulnerabilities, which lead to attack or incident.

An intrusion detection system is the prime solution, focused on preventing any attack from taking place. When considering any organizational infrastructure and its ICT components, there could be enormous loopholes (also known as vulnerabilities) that are left open to attack. System components that can be targeted for attack are categorized into physical components and digital components (Rao et al. 2016). Digital (or cyber) components include workstations, servers, routers, switches – all types of electronic network-connected devices. Physical components include power system, optics, and fibers, where there could be numerous vulnerabilities with the following attacks.

9.2.3 Attacks

Both cyber and physical components vulnerabilities can be found and targeted for attack. Any organization can face various attacks described below:

1. Cyber-physical attack (He & Yan 2016): This type of attack is done on interconnected installed devices, which can cause massive attack with associated vulnerabilities. This exposure of the structure was illustrated by the attack on the Ukraine power grid attack in 2016 (The Industrial Control Systems Cyber Emergency Response Team (ICS-CERT) 2016), in which the intruder introduced malware within the inter-connected communication channel, which in turn led the attacker to attain illicit access to the main control centers. Using this access, they extracted facts regarding critical lines of the grid in a confined region and launched an attack causing a wide-range power blackout.
2. Vandalism: In this type of attack the attacker causes deliberate damage or destruction of property, which could be public or private.
3. Interdiction attack (Salmeron, Wood, & Baldick 2004): This attack deals with the tripping of lines, transformers, generators in a grid.
4. Complex interconnected network-based attack (Cuadra et al. 2015): This type of attack is on the interconnected wires, whether it is communication system or electricity grids (Zhu et al. 2014; Hines, Dobson, & Rezaei 2016). In this, the attacker may have just an abstract knowledge of the structural connection or topology instead of the operational status; with this the attacker analyze the vulnerabilities of the system which could lead to series of attacks (Holme et al. 2002; Motter & Lai 2002; Rosas-Casals, Valverde, & Solé 2007; Wang & Rong 2009; Buldyrev et al. 2010).
5. Smart home attacks (Shouran, Ashari, & Priyambodo 2019): These attacks are on physical electronic smart gadgets that are inter-connected in a smart home and are more likely connected to the power system of the house. There could be vulnerability within the network when an intruder causes serious issues and it could affect both ways the external smart power grid as well as the house's smart meters and power supply.

6. Substation attack: This is one of the serious attack in which an invader probes into the deep layers of the firewall and attains full access to the substation (Ten, Liu, & Manimaran 2008). As the substation is compromised, the voltage control will become vulnerable to attacks. Even the configuration of the voltage controller could be manipulated, which in turn will lead to violations of voltage within the electric lines and oscillation within the system (Sridhar & Manimaran 2011).

Despite physical attacks on the entities, they cannot be successful without the support of cyber attacks as well, which are as follows:

1. Denial-of-service (DoS) and distributed denial-of-service (DDoS) attack (Carlin, Hammoudeh, & Aldabbas 2015): This is an attack in which numerous vulnerable computer systems attack a victim such as a server, website, or other network resource. For example, one major DDoS attack took place in Norway, where the attack stopped the online payment system of many big corporations (Wei 2014). Fig. 9.2 demonstrate an instance of Denial of service attack.
2. Man-in-the-middle (MitM) attack: This type of attack is nothing but eavesdropping on the conversation of two legitimate users, and it leads to leakage of the conversation as shown in Fig. 9.3.
    - Rouge Access Point: A wireless access point (AP) that is installed as a part of the target network. This AP is installed in a secure network with the help of a local administrator. This AP can be used to gain information by scanning for vulnerability in the network. This information can be used to attack the network using vulnerable points.
    - ARP Spoofing: This type of attack uses address resolution protocol (ARP) packets to deploy fake ARP packets in the network. The attacker tries to attach its MAC address to a legitimate IP address so that it can be the part of a secure network.

FIGURE 9.2
Denial of service.

- DNS Spoofing: Also known as DNS poisoning, it tries to introduce false DNS data by corrupting the DSN cache, resulting in an incorrect domain name-to-IP mapping. It can be used to redirect the user request to fake servers to capture sensitive user information.
- mDNS Spoofing: mDNS is similar to DNS but it's a multicast DNS in which each computer has its own list of DNS records.

Various techniques employed for the attack are as follows:

- Sniffing
- Packet injection
- Session hijacking
- Secure Socket layer(SSL) Striping

3. Phishing and spear phishing attacks (Giandomenico 2019): Both are online attacks, but they differ in the criteria of targeting the victim. Phishing attacks are targeted on a group of people so as to retrieve their personal information by fake links, whereas in spear phishing emails are targeted to a specific person who is fooled by the attacker to retrieve their personal information as shown in Figs. 9.4 and 9.5.
4. Drive-by attack: It is a common practice of spreading malware into the personal computer of the victim using http code. The victim gets trapped the moment they visit the website. They are also known as drive-by downloads attacks (Fig. 9.6).

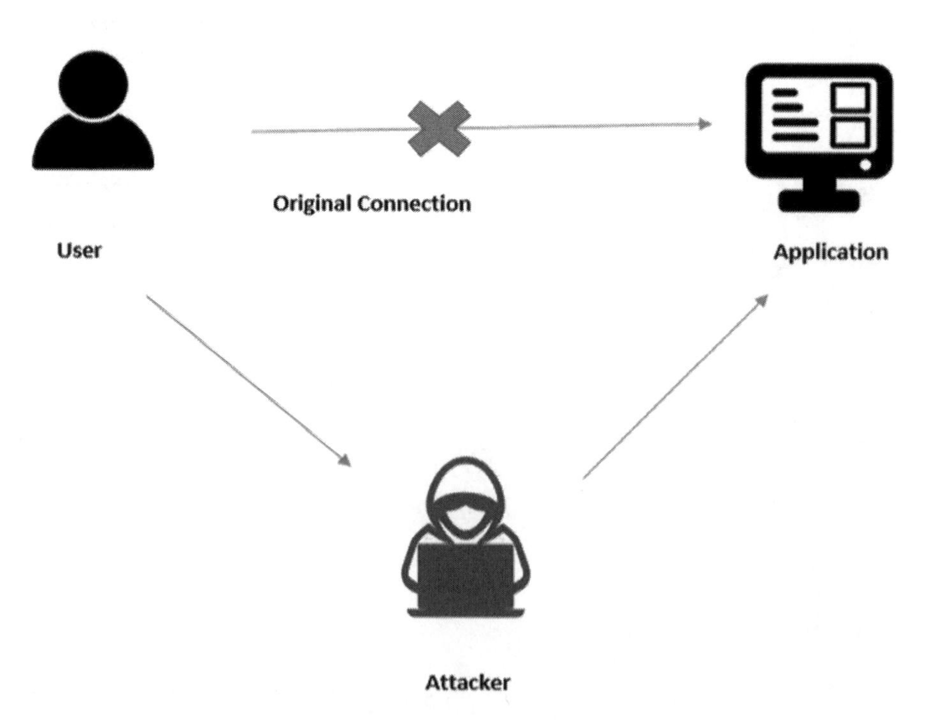

FIGURE 9.3
Man-in-the-middle attack.

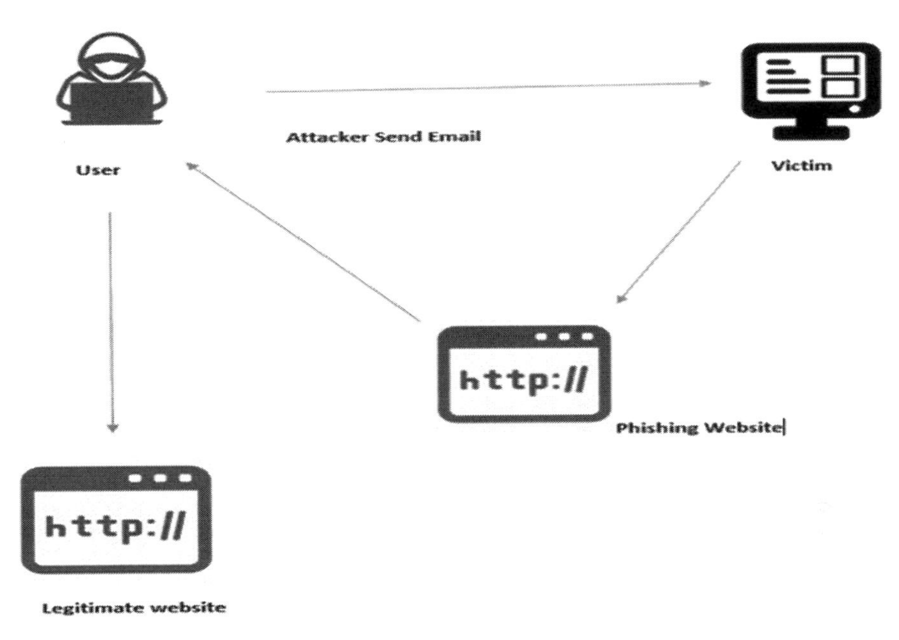

FIGURE 9.4
Phishing attack.

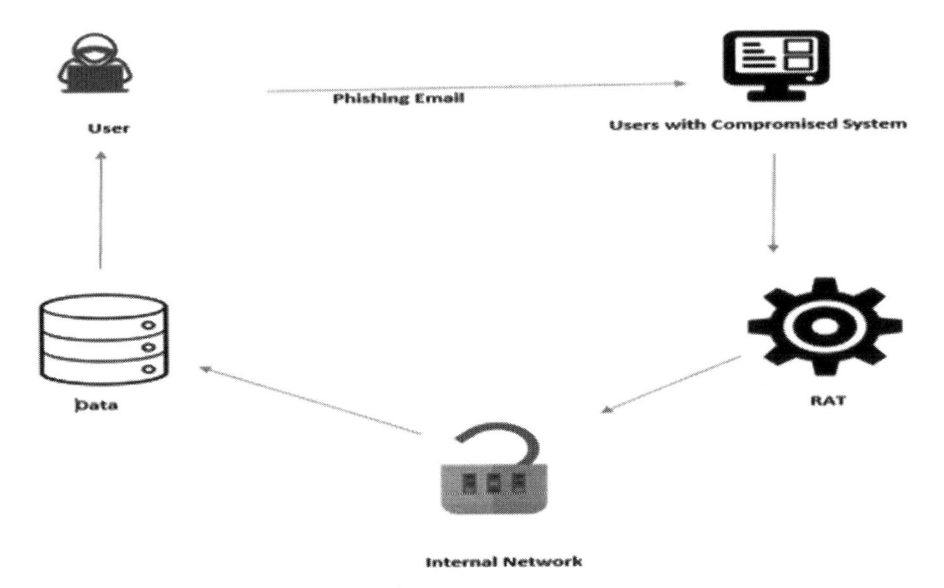

FIGURE 9.5
Spear phishing attack.

5. Password attack: It is actually a password-cracking methodology by which passwords are recovered by continuous guessing of the hash values of the target computer and setting the value in look up tables known as rainbow tables.
6. Structure Query Language (SQL) injection attack: This uses a code injection technique, inserted into the application with SQL statements that might carry malicious code within it. The depiction of the attack is shown in Figure 9.7.

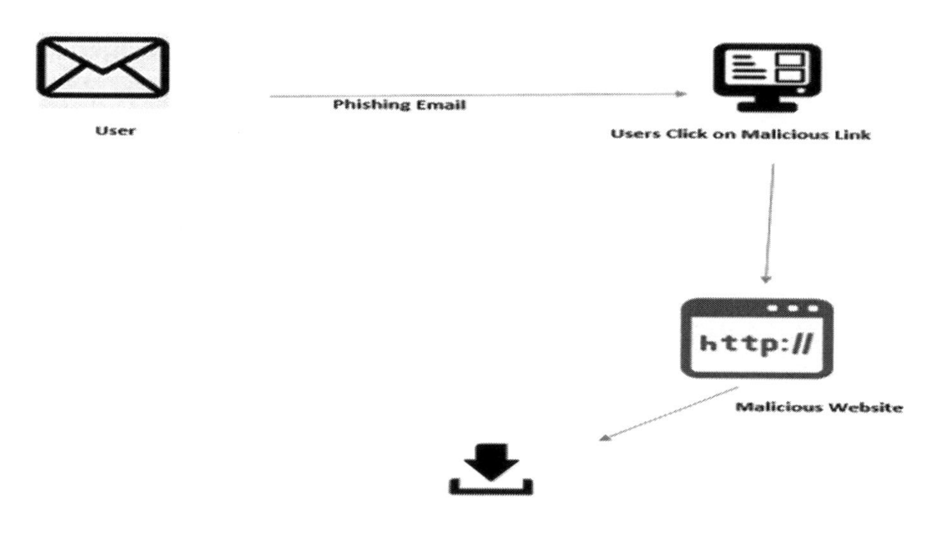

FIGURE 9.6
Drive-by attack.

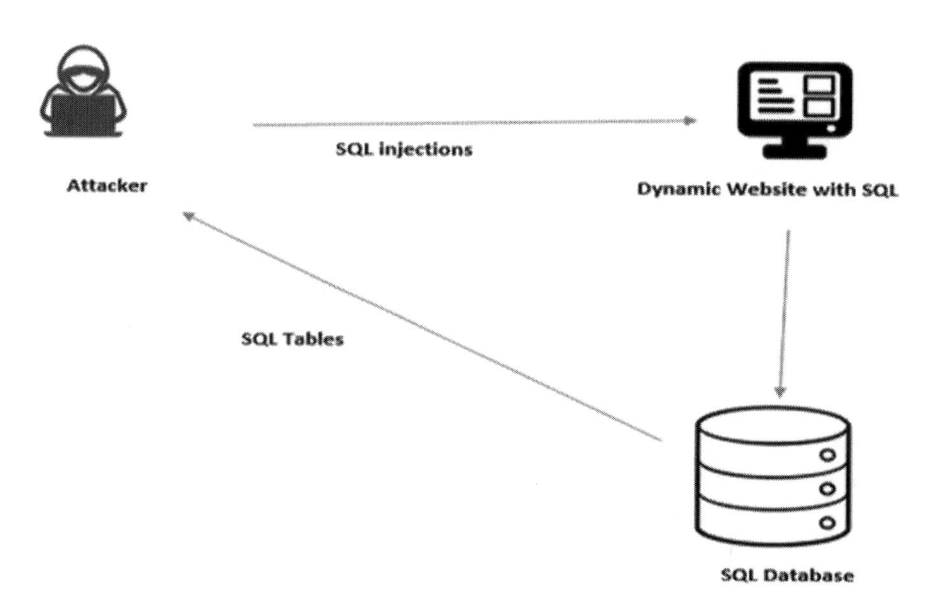

FIGURE 9.7
SQL injection.

7. Cross-site scripting (XSS) attack (Acunetix 2020): This too is an injection technique that happens to run on the client side of the system. In this scenario the nasty code is run in the web browser of the victim by hiding the malicious script in the web page or application, and the moment the user visits the web page or uses the application the attack occurs and executes the code as shown in Fig. 9.8.

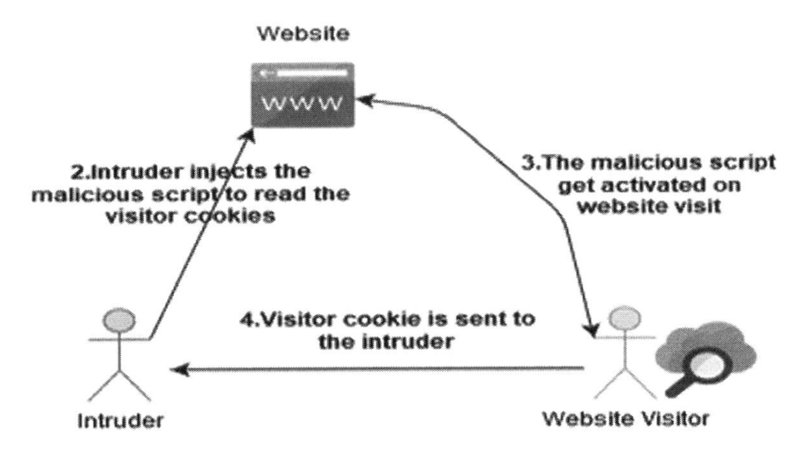

FIGURE 9.8
Cross-site scripting.

8. Eavesdropping (Frankenfield 2018): It is an attack in which the attacker seizes communication as shown in Fig. 9.9. It could be a sniffing or spoofing attack. Eavesdropping has been used as an attack several times. a few of them are as follows:
   - In 2011 an eavesdropping attack was targeted to android smartphones, using a sniffing program that involved the sending of authentication tokens over an unsecured WiFi network, which led to theft of the user's confidential data.
   - Another eavesdropping attack was carried out in 2015 when 25 000 iOS applications were affected because of a bug in the open-source library file AFNetworking, which led to the downfall of the HTTP encryption when the attacker amended the SSL by eavesdropping on the valid certificate.
9. Birthday attack: This is a cryptographic attack with an application of mathematic probability model. This model helps in resolving the complexities and know the applied hash function as shown in Fig. 9.10.

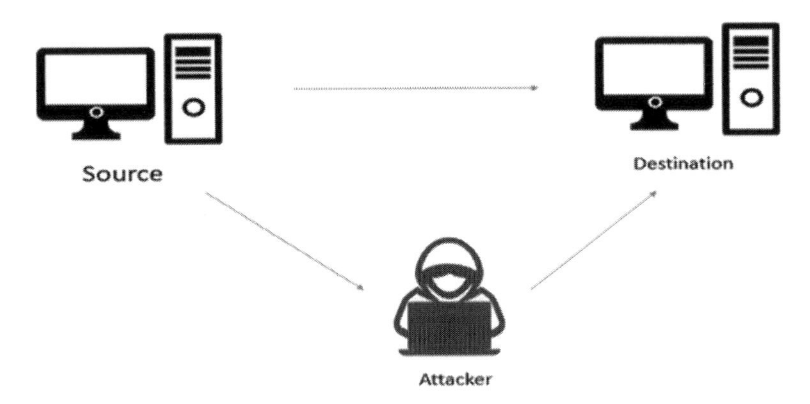

FIGURE 9.9
Eavesdropping attack.

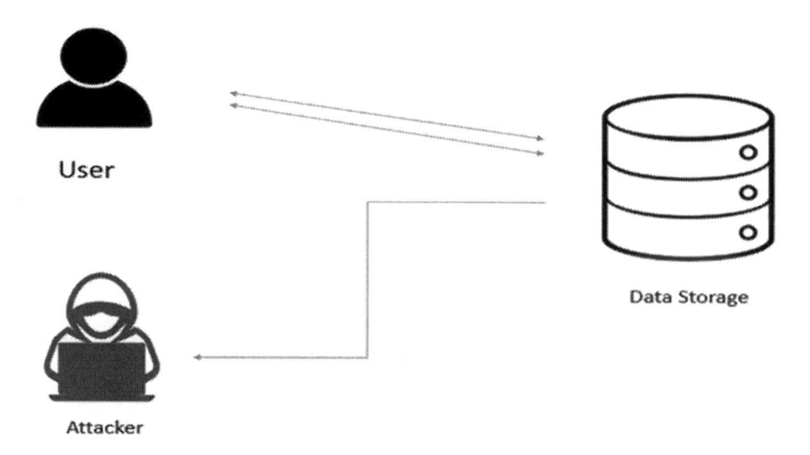

FIGURE 9.10
Birthday attack.

10. Malware attack: This is a cyber-attack in which malicious code is executed on the victim's computer without the user's knowledge as shown in Fig. 9.11. The various types of malwares are as follows:
    - Viruses
    - Worms
    - Trojans
    - Ransomware
    - Adware
    - Spyware
    - File-less malware
    - Hybrid attack

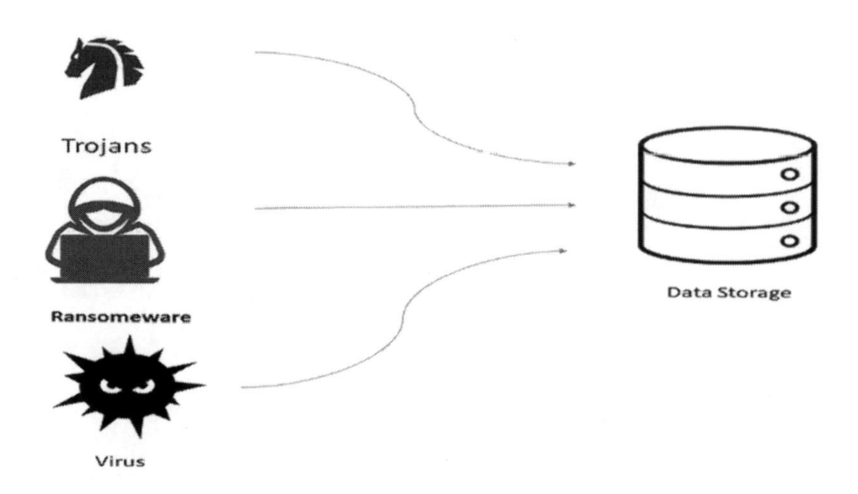

FIGURE 9.11
Malware attack.

The presence of malware in a system leads to various suspicious activities such as the following:

- CPU usage increases.
- Gadgets and browser speeds are affected and slow down.
- There are difficulties leading to network connectivity.
- There are sudden amendments in the files, programs, and icons.

To protect the system from such attacks a protective system is necessary.

9.3 Intrusion Detection System (Bace & Mell 2001)

The target of IDS is to check whether the system is working normally. Detection of difference from normal behavior indicates a possible attack or incident. An IDS find abnormal behavior by continuously monitoring and analyzing the usage pattern and network traffic for deviation from normal behavior as shown in Fig. 9.12.

Attack on a system is possible only when the system is compromised and the CIA is breached (Liao et al. 2013). In terms of both cyber and physical attacks there is need of an IDS that is robust in nature and reliable in functioning – a system that penetrates the network and finds the susceptible points.

9.3.1 IDS Architecture

IDS is analogous to an autonomous reviewing mechanism. It comprises three parts.

1. Agent: The agent resembles a logger who procures facts from the victim machine.
2. Director: The director resembles an analyzer who analyzes the flow of information and detects it as an attack in progress or intimates the occurrence of it.
3. Notifier: This analysis report is then passed onto the notifier, who decides how to notify the information as an essential entity.

9.3.2 Tasks Performed by IDS

IDS performs a certain set of tasks to prevent the attack, monitor user activities, perform network traffic analysis, track unusual alarms, and raise an alarm in the scenario of risk. The following tasks can be listed as the activities performed by the IDS:

1. Monitoring: The IDS continuously monitors user activities as well system behavior. There are certain set of rules, known as policies, that are defined for a user by the IT department of any organization. Every user is bound to follow those rules, and their activities are monitored for violation. A network log is captured to monitor the user activities. The software is known as a user activity monitor (UAM), which captures the user data to analyze for unusual activity. The UAM is used to prevent insider attacks. UAMs are generally part of the IDS or can be separately deployed. The system monitoring checks for CPU usage, memory usage, and other resource usage.

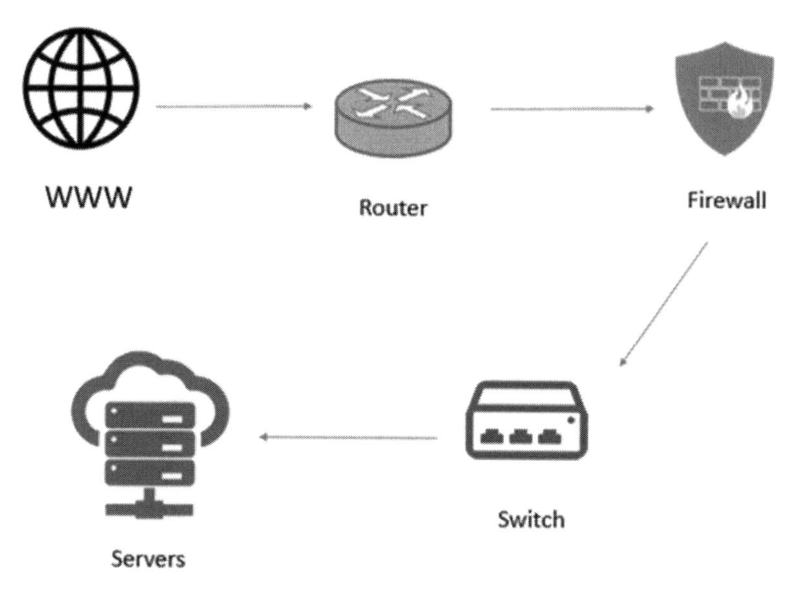

FIGURE 9.12
Intrusion-detection system in a network.

2. Analysis: User data as well as the system data are continuously under analysis. The analysis techniques usually employ pattern detection techniques of machine learning. These patterns are matched against the pattern of an attack to check if the current state of the system in under threat or an incident as already occurred. If the system is under threat the IDS should raise an alarm so that preventive measure be taken by the IT.

   The IDS also analyzes the system configuration and vulnerabilities of the system. The system configuration is checked for any kind of wrong or unsafe configuration (such as unused open port, inconsistent access control list (ACL)) which can become vulnerabilities of the system.
3. Integrity check: The IDS continuously checks for integrity of the system and files to detect if they are compromised. Apart from access control for file, an integrity check is used to ensure that the file data has not been modified. An integrity check is used as a part of IDS and it uses hash values to match if the file is modified or not. Integrity check updates hash values after legitimate modification of the file and regularly matches the hash values with the file hash to check the integrity.
4. Pattern extraction and matching: Pattern includes the user behavior pattern, resource usage pattern, and network traffic. By employing various techniques (such as statistical or ML) the pattern can be extracted from these data. The pattern of normal behavior is matched with the system behavior and analyzed for the abnormality.
5. Policy enforcement: Every organization defines its policies to implement access control, security, authentication, and data protection. The IDS always monitors the user to adhere to the policy, and any violation will be blocked and reported by the IDS.

9.3.3 IDS Categorization

IDS could be categorized on the following basis (Liao et al. 2013).

1. Detection methodologies of the IDS
2. Detection approaches of IDS
3. Target environment of IDS

The categorization is demonstrated in figure 9.13.

9.3.3.1 Detection Methodologies of the IDS (Bace & Mell 2001)

a. Signature-based detection: It is also known as knowledge-based or misuse detection in which patterns are compared to detect an intrusion. Signatures are treated as the presence of an attack, and if the pattern matches with the signature of an attack the IDS assumes and raises an alarm to the security architecture.
b. Anomaly-based detection: Another name of this technique is behavior-based detection in which an intrusion is detected by a change in behavior such that normal profiles are equated with the occurring event. Anomaly detection works on observing the resource usage and user behavior. Normal behavior is defined by processing the data by some ML techniques and stored. If continuous analysis of the usage and behavior doesn't fit in the trained model classes, the behavior is assumed as an anomaly and the corresponding alarm will be raised by the IDS.
c. Stateful protocol analysis: This analysis tracks the protocol state of the communication system, for example, when there is a request of paring for communication.

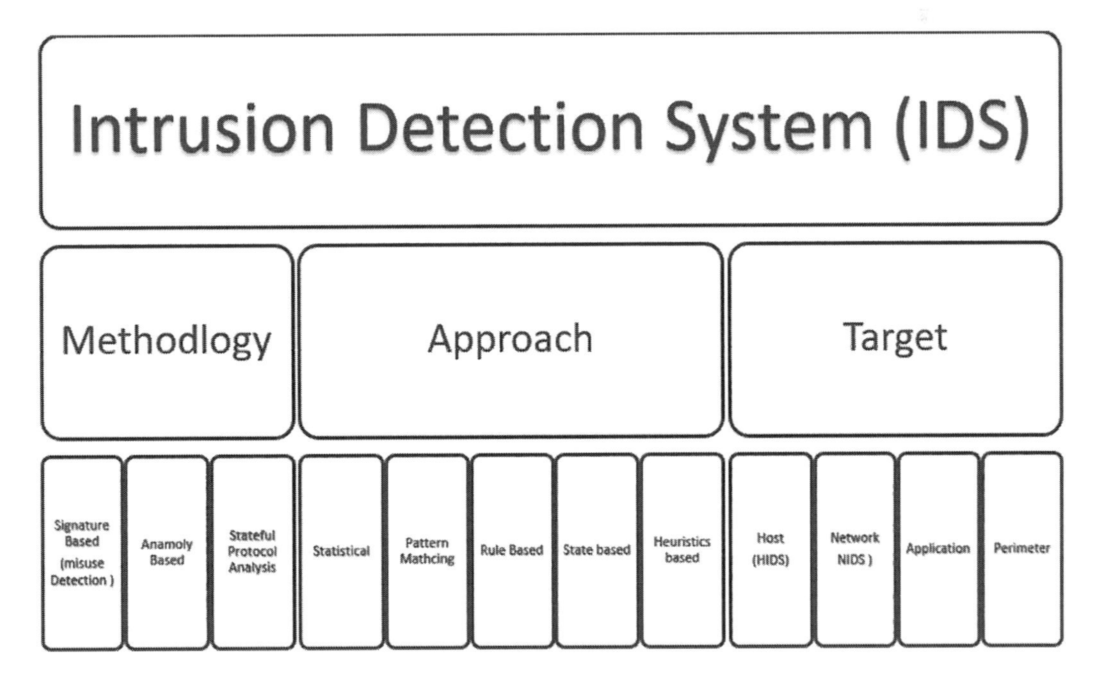

FIGURE 9.13
Classification of IDS.

9.3.3.2 Detection Approaches

As summarized in table 9.1, IDS categorization is based on intrusion detection technique:

a. Statistics-based
b. Pattern-based
c. Rule-based
d. State-based
e. Heuristic-based

9.3.3.3 Categorization of the IDS Based on the Target Environment

a. Host-based IDS: The target of host-based IDS (HIDS) is the individual computer, which is monitored and analyzed for the intrusion. HIDS runs on the individual system. It uses the support and information of the operating system and analyzes the user behavior data to check for intrusion.
b. Network-based IDS: The network intrusion detection (NIDS) is an intrusion detection system that works on monitoring and analyzing the network traffic and

TABLE 9.1

Classifications and Comparisons of Various Intrusion Detection Approaches

S. No.	Detection Approach	Technology Type	Detection of Attack	Performance	Type of Sources	Other Characteristics
1	Statistics-based [24]	Host-based IDS & network-based IDS	Both known and unknown attacks	Medium	Network packet traffic Audit data	Simple but less accurate
2	Pattern-based	Host-based IDS & network-based IDS	Known attacks	High	User-defined known intrusion signatures	Notifying users typing pattern Checking of file integrity Simple but less flexible
3	Rule-based	Host-based IDS & network-based IDS	Both known and unknown attacks	Medium	Audit data user profiles Network packets AP profile	Higher accuracy and flexibility
4	State-based	Host-based IDS & network-based IDS	Unknown attacks	High	Transition diagram of known attacks Sequence of system calls or commands	Probability-based self-learning
5	Heuristic-based	Network-based IDS	Both known and unknown attacks	Medium	Audit records Network traffic	

individual packets to check for anomalies. The content and format of the individual packets are checked to find the intrusion.

c. Application-based IDS: This is an IDS system that is installed specifically for analyzing and monitoring specific applications within a system.
d. Perimeter-based intrusion detection system: This is an IDS system that is designed to protect assets within a defined area.

9.4 Intrusion Detection Using Machine Learning

9.4.1 Machine Learning

Machine learning is a branch of artificial intelligence, which provides the capability to the machine (computational engine) to acquire knowledge from learning the system and environment without the need of hardcode program knowledge. ML is learning by experience (i.e., the machine learns from observations and data, and the acquired knowledge will be used for decision making in classification, clustering, etc.).

Traditional methods of IDS are not suitable for today's dynamic nature and high rate of data production, so newer implementation and techniques are required for modern IDSs. By use of machine learning techniques, the dynamic nature and rate of data can be managed by using adaptive learning algorithms, which are also capable of detecting new attacks. The advantages of ML in IDS are that it provides high response time, is adaptable, is scalable, and gets fewer false positives.

The task of ML-based IDS is a simple classification task with two classes. One is malicious behavior, and other is non malicious behavior. Machine learning approaches for IDS can be divided into two categories (Zamani & Movahedi 2013). It is diagrammatically represented in figure 9.14.

9.4.2 Artificial Intelligence-Based Approaches

(Laskov et al. 2005). This kind of approach traditionally includes statistical and mathematical modeling. The AI techniques include k-nearest neighbor (k-NN), support vector machine (SVM), and multi-layer perceptron (MLP).

9.4.2.1 K-NN

K-nearest neighbor is a supervised learning algorithm based on feature similarity and classification performed by classifying the data points based on how the neighbors are classified. It is performed using a k-nn classifier. To match the features accurately, the K factor parameter is tuned. The value of K is selected by two methods:

a. Sqt(n), where n is the total number of data points.
b. In case there are even values, odd values of K are selected to avoid confusion between the classes of data.

K-NN is used in scenarios in which data sets are small, noise free, and labelled.

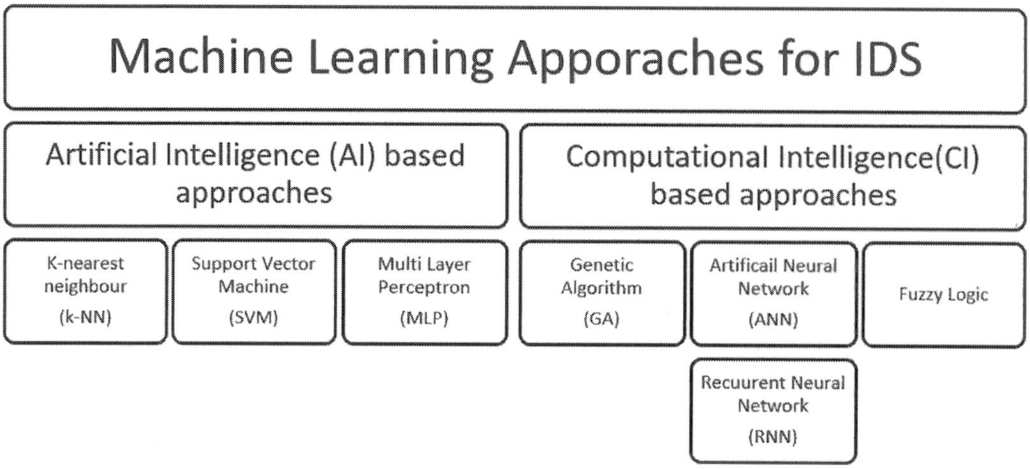

FIGURE 9.14
Machine learning–based approaches for IDS.

9.4.2.2 SVM

A support vector machine is a supervised learning algorithm, which classifies the objects belonging to different classes. This differentiation is done by drawing a hyperplane. This hyperplane is drawn when data points are plotted with respect to the z-axis. There are complexities in defining the classes of the data sets when there are overlapping data points. To perform this classification there are two parameters such as

a. regularizing parameter
b. gama

To achieve accurate values in a specified time, the values of theses parameters can be tuned in an SVM classifier

9.4.2.3 MLP

Multilayer perceptron is one of the feeds forward artificial neural networks, which uses backpropagation, for training purposes. A multilayer perceptron has multiple layers that use nonlinear activation functions to give the desired output. A multilayer perceptron has layers such as input layer, output layer and hidden layer. The MLP inner layer takes the input; the hidden layer manipulates the weights; and the output layer displays the results. The backpropagation training is necessary because the weights get updated by the output layer towards the input layer.

9.4.3 Computational Intelligence-Based Approaches

This kind of approach includes an evolutionary algorithm, artificial neural network, and fuzzy logic.

9.4.3.1 Genetic Algorithm (GA)

A genetic algorithm (Mallawaarachchi 2017) is an exploratory that was stimulated by Charles Darwin's theory of natural evolution. This algorithm reveals the method of natural assortment in which apt individuals are designated for reproduction to produce offspring for the upcoming generation. In this algorithm at every step, the current population generates children by selecting individuals known as parents who provide their genes for the entries of their vectors – to their children. There are five phases of genetic algorithm:

1. Initial population
2. Fitness function
3. Selection
4. Crossover
5. Mutation

9.4.3.2 Artificial Neural Network (ANN)

An artificial Newual Network (Bivens et al. 2002) is a calculative model inspired by structural and functional aspects of brain. It is a statistical model in which the multifaceted relationship between inputs and outputs are patterned in a model. The biological neuron present in the brain is artificially developed, and various models are being developed to provide the desired output. An artificial neural network can accomplish tasks that linear programs cannot perform. Despite reprogramming, a neural network can learn and can handle the missing data. Neural networks usually implement supervised learning tasks, which means structuring knowledge from data sets in terms of the known output in advance. The ANN then trains itself and deduces the correct answer on its own.

9.4.3.3 Fuzzy Logic

Fuzzy logic is a multi-valued logic in which true values of variables may be any real number between 0 and 1 both inclusive. It is engaged to manage the theory of partial truth, where the true value may be among completely true and completely false (Singh et al. 2013). There are various applications where fuzzy logic is used, such as facial configuration and recognition, air conditioners automation and sensing, automatic washing machines, smart vacuum cleaners, antiskid braking systems in cars, transmission systems, unmanned vehicles, optimization of power systems in smart cities and smart homes, weather forecasting systems, predictive models for upcoming new products in market and their pricing, prediction on project risk assessment, medical diagnosis and predicting treatment plans, and stock trading. It was intended to permit the computer to decide and distinguish among data that could be either true or false, similar to human reasoning.

9.4.4 Computer Forensics

Computer forensics is one of the methods used to scrutinize and analyze various techniques to assemble and protect evidence from the victim's device in a documented format so that it is presentable in the court of law.

A digital footprint or digital shadow represents a unique set of trackable digital events, movements, assistances, and infrastructures exhibited on the internet or on digital devices.

Digital evidence is defined as facts and data that is used to explore the stored, received, or transmitted data by electronic gadgets. This evidence could be extracted from the seized gadgets and examined.

9.4.5 Pattern Recognition

To manually analyze the design patterns of data to detect a security threat is a tedious job to accomplish. To overcome the effort, machine learning methods offer models that analyze the data and discover the potential threat. Intrusion detection can be thought of as a task of pattern recognition form the network log or user log. ML techniques are broadly classified as supervised learning, unsupervised learning, and reinforced learning. Pattern recognition can be done with both supervised leaning and unsupervised learning.

Intrusion detection systems apply machine-learning methodologies, which can be further divided into two categories:

1. supervised
2. unsupervised (Laskov et al. 2005)

9.5 Machine Learning Approaches in WSN Security: Case Studies

Wireless networks are one of the most significant ones as they possess the property of being remotely operable with the help of smart and efficient sensor nodes. With incomparable properties of being intelligent, self-consolidating, and self-restorative, the network still has many shortcomings, such as short computational ability, limited storage, and limited battery life (Baraneetharan 2020), which makes it susceptible to attacks. Machine learning approaches perform a key role in the IDS of WSN. In a WSN the location of the data specifies the deployment of the IDS, which is centralized IDS or distributed IDS. The theft of data from within the network specifies the type of IDS to be used such as network-based IDS, host-based IDS, and hybrid-based IDS.

Machine learning approaches used in WSN are broadly categorized as follows:

1. Anomaly-based detection approach
2. Misuse-based detection approach
3. Hybrid-based detection approach
4. Clustering-based detection approach
5. Trust-based detection approach

Based on the above approaches, several research projects have been done.

9.5.1 Anomaly-Based Detection Approach

Anomalies exists in every scenario. In the case of WSNs, many defects or irregularities can occur at different levels, such as remotely deployed sensor nodes, network connection within devices, irregularity of data flow, or loss of network due to fluctuating signals. All anomalies give a loophole to the attacker. Such problems generally could be because of environmental conditions or other technical issues; at the sensor node level they could be because of irregularity of power due to failure of solar panels (Islam & Rahman 2011). Many intrusion techniques use this approach, such as statistical models, clustering algorithms, artificial immune systems, isolation tables, and game theory (Chen, Hsieh, & Huang 2009; Soliman, Hikal, & Sakr 2012; Abduvaliyev et al. 2013; Tan et al. 2019), which employ rule-based methods to detect the intrusion, where specific models are defined for the regular activity. Any aberration from that is termed an anomaly and an intrusion is detected.

9.5.2 Misuse-Based Detection Approach

This approach works on matching pattern, which are known as "signatures." The patterns of known attacks are given to the IDS in some format so as to match and stop the incoming intrusion. But the major disadvantage of this approach is that it cannot identify any new attack as it does not have any predefined rules (Hai, Khan, & Huh 2007). A genetic algorithm category of colonial selection uses a watchdog algorithm to detect intrusion. It continuously identifies and distinguishes the nodes for anomalous conduct in the process of data transmission while keeping an eye on the adjoining nodes within a network. This adversely affects the working of the WSN. To compensate for this problem, Destination-Sequenced Distance vector Routing algorithm is also used, which continuously updates the routing table, which reduces the load on the remote nodes. One such approach is shown by Adil et al. (2020) in their research, where a secure reliable sensor network is developed for wirelessly charging the electric vehicle. They enhanced the DSDV using machine learning approach.

9.5.3 Hybrid-Based Detection Approach

The network administrator defines the security protocols in a WSN. This approach is a combination of both anomaly and misuse techniques. The usage of this approach is well demonstrated by Sedjelmaci and Feham (2011). In their work an IDS is developed for clustered wireless sensor network using hybrid approach in which anomaly detection based on SVM and misuse are combined. Their system detected most of the routing attacks. Wireless networks having characteristics of open-air transmission and with no fixed infrastructure are more susceptible to attack. To save them (Maleh et al. 2015) has proposed an IDS whose basic approach is to save energy within the network and unburden the sensor nodes.

9.5.4 Clustering-Based Detection Approach

If continuous monitoring is not done within a network, many resources are wasted (Wang, Attebury, & Ramamurthy 2006). To counter this problem IDS based on clusters was developed in which each node within the cluster monitors the intrusion attack and its source. In a few scenarios, Mobile Ad hoc network (MANET) is also used as centralized IDS. Research has been done to boost the detection of the sensor nodes and lessen the false rate (Sun et al. 2015). To detect an anomaly of the sensor node, head node, or sink node AdaBoost algorithms with hierarchical structure are employed. To detect the distorted sink nodes backpropagation is optimized and used by cultural algorithm and artificial fish swarm algorithm.

9.5.5 Trust-Based Detection

This is also known as reputation-based IDS where each node is graded based on the cooperation and contribution of a node within a network. This is mostly done to make a trusted member network based on the reputation. It is categorized in three types as follows:

- Subjective reputations are evaluated based on the direct neighbor interactions.
- Indirect reputations are evaluated based on non-neighbor and network interactions.
- Functional reputations are based on both subjective and indirect interactions.

Using all three, a reputation-specific chart is built that is analyzed to stop the upcoming attack. In such scenario, Dynamic Source Routing Protocol (DSR) is used, which rates the nodes based on performance. The watchdog algorithm identifies any doubtful action on the node in the main route, alarms the other in the network, and calculates the reputation of the node.

9.6 Future Aspects and Conclusion

WSN being an intelligent network has many issues that have been solved by machine learning, but still new security efforts related to the research in this field are still awaited. Areas such as recognition precision through sensors so as to conserve the energy in transmission of data and extend the life of the network can be achieved using efficient algorithms of machine learning. With the changing environment of resources and its needs, the nodes should self-adapt themselves as per the requirement of the network. This could be achieved using machine learning techniques.

References

Abduvaliyev, A., Pathan, A. S. K., Zhou, J., Roman, R., & Wong, W. C. (2013). On the vital areas of intrusion detection systems in wireless sensor networks. *IEEE Communications Surveys & Tutorials*, *15*(3), 1223–1237.

Acunetix (2020). Cross site scripting XSS. Retrieved from https://www.acunetix.com/websitesecurity/cross-site-scripting/

Adil, M., Ali, J., Ta, Q. T. H., Attique, M., & Chung, T. S. (2020). A reliable sensor network infrastructure for electric vehicles to enable dynamic wireless charging based on machine learning technique. *IEEE Access*, *8*, 187933–187947.

Bace, R.G., & Mell, P. (2001). Intrusion detection systems. http://csrc.nist.gov/publications/nistpubs/800-31/sp800-3 1.pdf

Baraneetharan, E. (2020). Role of machine learning algorithms intrusion detection in WSNs: a survey. *Journal of Information Technology*, 2(03), 161–173.

Bivens, A., Palagiri, C., Smith, R., Szymanski, B., & Embrechts, M. (2002). Network-based intrusion detection using neural networks. *Intelligent Engineering Systems through Artificial Neural Networks*, *12*(1), 579–584.

Buldyrev, S. V., Parshani, R., Paul, G., Stanley, H. E., & Havlin, S. (2010). Catastrophic cascade of failures in interdependent networks. *Nature*, *464*(7291), 1025–1028

Carlin, A., Hammoudeh, M., & Aldabbas, O. (2015). Defence for distributed denial of service attacks in cloud computing. *Procedia computer science*, *73*, 490–497

Chen, R. C., Hsieh, C. F., & Huang, Y. F. (2009, February). A new method for intrusion detection On hierarchical wireless sensor networks. In *Proceedings of the 3rd International Conference on Ubiquitous Information Management and Communication* (pp. 238–245).

Crume, J. (2000). *Inside Internet Security: What Hackers don't Want You to Know*. Addison-Wesley, United Kingdom.

Cuadra, L., Salcedo-Sanz, S., Del Ser, J., Jiménez-Fernández, S., & Geem, Z. W. (2015). A critical review of robustness in power grids using complex networks concepts. *Energies*, *8*(9), 9211–9265.

Frankenfield, J. (2018, Mar 10). Eavesdropping Attack Defined retrieved from https://www.investopedia.com/terms/e/eavesdropping-attack.asp

Giandomenico, N. (2019, October 24). What is Spear-phishing? Defining and Differentiating Spear-phishing from Phishing. Retrieved from https://digitalguardian.com/blog/what-is-spear-phishing-defining-anddifferentiating- spear-phishing-and-phishing

Hai, T. H., Khan, F., & Huh, E. N. (2007, August). Hybrid intrusion detection system for wireless sensor networks. In *International Conference on Computational Science and Its Applications* (pp. 383–396). Springer, Berlin, Heidelberg.

He, H., & Yan, J. (2016). Cyber-physical attacks and defences in the smart grid: a survey. *IET Cyber-Physical Systems: Theory & Applications*, *1*(1), 13–27

Hines, P. D., Dobson, I., & Rezaei, P. (2016). Cascading power outages propagate locally in an influence graph that is not the actual grid topology. *IEEE Transactions on Power Systems*, *32*(2), 958–967.

Holme, P., Kim, B. J., Yoon, C. N., & Han, S. K. (2002). Attack vulnerability of complex networks. *Physical review E*, *65*(5), 056109

Islam, M. S., & Rahman, S. A. (2011). Anomaly intrusion detection system in wireless sensor networks: security threats and existing approaches. *International Journal of Advanced Science and Technology*, *36*(1), 1–8.

Laskov, P., Düssel, P., Schäfer, C., & Rieck, K. (2005, September). Learning intrusion detection: Supervised or unsupervised?. In *International Conference on Image Analysis and Processing* (pp. 50–57). Springer, Berlin, Heidelberg

Liao, H. J., Lin, C. H. R., Lin, Y. C., & Tung, K. Y. (2013). Intrusion detection system: a comprehensive review. *Journal of Network and Computer Applications*, *36*(1), 16–24.

Maleh, Y., Ezzati, A., Qasmaoui, Y., & Mbida, M. (2015). A global hybrid intrusion detection system for wireless sensor networks. *Procedia Computer Science*, *52*, 1047–1052.

Mallawaarachchi, V. (2017, Jul 8). Introduction to Genetic Algorithms — Including Example Code. Retrieved from https://towardsdatascience.com/introduction-to-genetic-algorithms-including- example-code-e396e98d8bf3

Motter, A. E., & Lai, Y. C. (2002). Cascade-based attacks on complex networks. *Physical Review E*, *66*(6), 065102.

Raj, A. B., Ramesh, M. V., Kulkarni, R. V., & Hemalatha, T. (2012, June). Security enhancement in wireless sensor networks using machine learning. In *2012 IEEE 14th International Conference on High Performance Computing and Communication & 2012 IEEE 9th International Conference on Embedded Software and Systems* (pp. 1264–1269). IEEE.

Rao, N. S., Poole, S. W., Ma, C. Y., He, F., Zhuang, J., & Yau, D. K. (2016). Defense of cyber infrastructures against cyber-physical attacks using game-theoretic models. *Risk Analysis*, *36*(4),694–710.

Rosas-Casals, M., Valverde, S., & Solé, R. V. (2007). Topological vulnerability of the European power gridunder errors and attacks. *International Journal of Bifurcation and Chaos*, *17*(07), 2465–2475

Salmeron, J., Wood, K., & Baldick, R. (2004). Analysis of electric grid security under terrorist threat. *IEE Transactions on power systems*, *19*(2), 905–912.

Sedjelmaci, H., & Feham, M. (2011). Novel hybrid intrusion detection system for clustered wireless sensor network. *arXiv preprint arXiv:1108.2656*.

Shouran, Z., Ashari, A., & Priyambodo, T. K. (2019). Internet of things (IoT) of smart home: privacy and security. *International Journal of Computer Applications*, *182*(39), 3–8.

Singh, H., Gupta, M. M., Meitzler, T., Hou, Z. G., Garg, K. K., Solo, A. M., & Zadeh, L. A. (2013). Real-life applications of fuzzy logic. *Advances in Fuzzy Systems, 2013*, 1687–7101.

Soliman, H. H., Hikal, N. A., & Sakr, N. A. (2012). A comparative performance evaluation of intrusion detection techniques for hierarchical wireless sensor networks. *Egyptian Informatics Journal, 13*(3), 225–238.

Sridhar, S., & Manimaran, G. (2011, July). Data integrity attack and its impacts on voltage control loop in power grid. In *2011 IEEE Power and Energy Society General Meeting* (pp. 1–6). IEEE.

Stallings, W. (2006). *Cryptography and Network Security, 4/E*. Pearson Education, India.

Sun, X., Yan, B., Zhang, X., & Rong, C. (2015). An integrated intrusion detection model of cluster-based wireless sensor network. *PloS One, 10*(10), e0139513.

Tan, X., Su, S., Huang, Z., Guo, X., Zuo, Z., Sun, X., & Li, L. (2019). Wireless sensor networks Intrusion detection based on SMOTE and the random forest algorithm. *Sensors, 19*(1), 203.

Ten, C. W., Liu, C. C., & Manimaran, G. (2008). Vulnerability assessment of cybersecurity for SCADA systems. *IEEE Transactions on Power Systems, 23*(4), 1836–1846.

The Industrial Control Systems Cyber Emergency Response Team (ICS-CERT) (2016). 'Cyber-attack against Ukrainian critical infrastructure'. Alert (IR-ALERT-H-16-056-01). Available at: https://www.ics-cert.us-cert.gov/alerts/IR-ALERT-H-16-056-0

Wei, W. (2014, May 14, 2015). Sony Playstation Network Taken Down by DDoS Attack. The Hackers News. Available: http://thehackernews.com/2014/08/sony-playstation-network-taken-down-by_24.html

Wang, J. W., & Rong, L. L. (2009). Cascade-based attack vulnerability on the US power grid. *Safety Science, 47*(10), 1332–1336

Wang, Y., Attebury, G., & Ramamurthy, B. (2006). A survey of security issues in wireless sensor networks. *IEEE Communications Surveys & Tutorials.*

Zamani, M., & Movahedi, M. (2013). Machine learning techniques for intrusion detection. *arXiv preprint arXiv:1312.2177.*

Zhu, Y., Yan, J., Sun, Y. L., & He, H. (2014). Revealing cascading failure vulnerability in power grids using risk-graph. *IEEE Transactions on Parallel and Distributed Systems, 25*(12), 3274–3284.

10

Security in Communication for Intelligent Wireless Sensor Networks
Issues and Challenges

Shiwangi Singh, Ritwik Saurabh, Tanmoy Maitra, and Debasis Giri

CONTENTS

DOI: 10.1201/9781003102397-10

10.1 Introduction

Due to the recent advances in very small-scale processing and the high-scale incorporation of multiple electronic components into a single package, wireless sensor networks are an evolving technology. The standard sensor node is a stand-alone electronic packet that is needed to accommodate a variety of sensors, an embedded microcontroller, a limited-capability device that can be renewable or not, and a radio transceiver in its core. The standard sizes of a sensor node are from a matchbox to a coin, but the thrilling promise of nanotech development and processing is to shrink the size drastically over the next ten years.

In order to develop intelligent applications in the future, artificial intelligence and wireless sensor networks are coordinated to answer numerous questions including computing intelligence, wireless sensor networking, artificial intelligence, neural networks, radio-frequency (RF) recognition, IoT, computation parallel, autonomous driving, machine learning (ML) and global positioning system (GPS) technology.

The organization of this chapter is as follows: Section 10.2 explains the layer-based architecture. Section 10.3 briefly describes security aspect of communication. Section 10.4 compares the layer-based secure communication protocols. Possible attacks in communication are highlighted under Section 10.5. Proceeding with Section 10.6, issues and challenges in communication protocols are demonstrated. Finally, a conclusion with a future scope is given in Section 10.7.

10.2 Layer-Based Architecture

A few hundred sensor nodes and a single strong base station are used in layered network architecture. Concentric layers of network nodes are coordinated. It is made up of five layers and three cross layers (See Figure 10.1).

10.2.1 Radio Layer

Data transmission via radio frequency signals through the air is under the responsibility of the radio layer. It protects the transmitting medium.

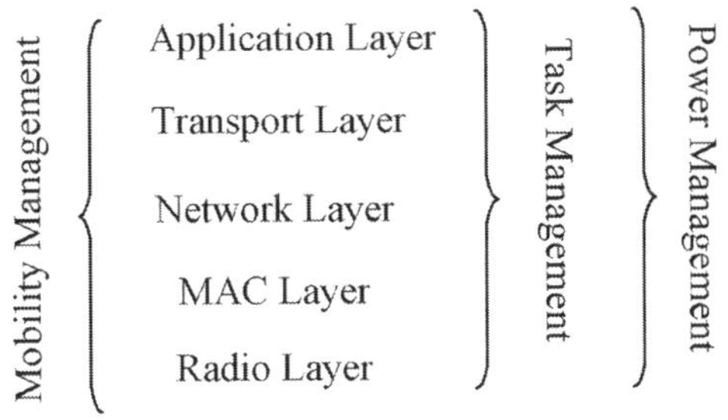

Mobility Management
Application Layer
Transport Layer
Network Layer
MAC Layer
Radio Layer
Task Management
Power Management

FIGURE 10.1
Layer-based architecture.

10.2.2 MAC Layer

The medium-access control (MAC) layer gives the Logical Link Control (LLC) and high layers of the open system interconnect (OSI) network an abstraction of the physical layer. One of the main responsibilities of this layer is the frame-encapsulation so that the frames can be transmitted across a physical communication network. It resolves the address of the source station, target group stations, or destination station. It defines the channel entry methods for transmission. It also resolves collisions and retransmits when collisions occur. Different categories of MAC layer protocols are shown in Figure 10.2.

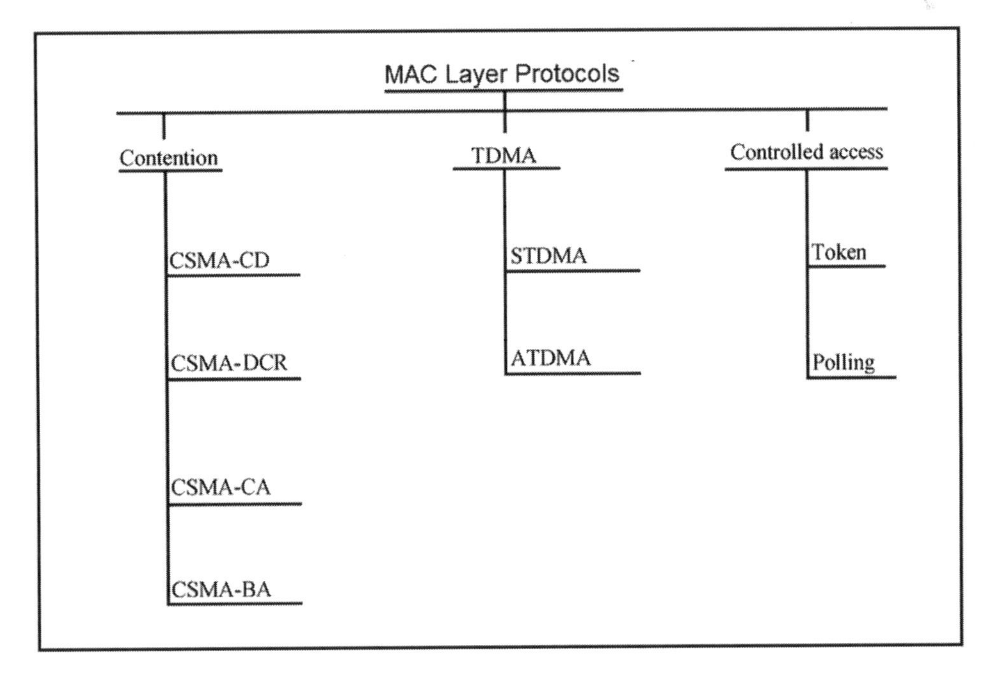

FIGURE 10.2
Categories of MAC layer protocols.

10.2.3 Network Layer

Routing is a major feature of the network layer. It has numerous applications-based functions, but in practice it has no universal identification and must be self-organized. The key tasks are power retention, partial memory, buffers, and sensors.

The concept behind the routing protocol is simply to illustrate a reliance on a stable line and redundant lanes, using a persuaded metric scale. Many existing protocols can be divided into this network layer: flat and hierarchical routing and possibly divided in a time-based event and query driven protocol [1].

Figure 10.3 shows the working functionality of the radio and MAC layers with respect to data transmission from one network to another.

10.2.4 Transport Layer

Congestion avoidance and reliability are the main functions of the transport layer. These protocols use various methods to detect and restore damage. If a device is intended for communication with other networks, the transport layer is required.

Providing a more energy-efficient stable loss recovery is one of the key reasons that the Transmission Control Protocol (TCP) is inadequate for WSN. Transportation layers can typically be divided into event-based and packet-driven. There are several transport layer protocols which are used in a variety of networks like Stream Control Transmission Protocol (SCTP), Pump Slowly Fetch-Quickly (PSFQ), etc. There are other protocols available in this layer [2].

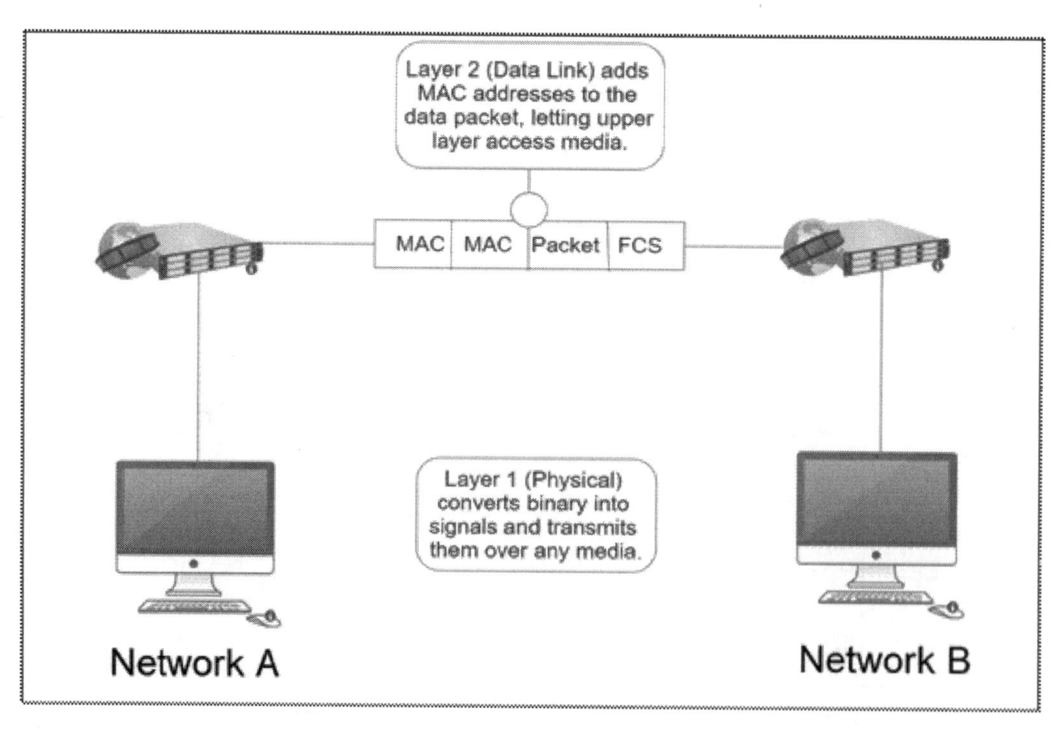

FIGURE 10.3
Working functionality of the radio and MAC layer.

10.3 Security in Communication

Secure communication is where several people interact and do not expect a third person to listen. That is why they ought to talk in a manner that is not vulnerable to attract others. There can be much vulnerability in communication, which can be broadly classified as follows.

10.3.1 IoE Vulnerabilities

Although the major advantage of edge computing is that it fixes a variety of network problems in an internet of everything (IoE) network, it may introduce fresh susceptibilities, as a result of which the total threat surface (total device access points that a potential attacker may strike) increases. At both the edge and at the consumer endpoint, the state of present systems makes communication very susceptible. For example, many attacks can arise making users unable to alter previous passwords, allowing attackers to access the network, and also exposing edge computers for vulnerabilities.

Dangerous internet services could be easily located and downloaded. (A search engine was released around 2013 to discover unsafe IoE points round the globe.) Five thousand IoT computers have been targeted by 5,000 distinct programs attempting to hack into vulnerable or default codes.

Although vulnerability to such attacks stays at the terminal, edge computing increases difficulties by incorporating possible fresh target surfaces. IoE applications that establish a connection to a public network violate authentication protocols specifically from edge of the network. This is partially due to the present state of edge computing in which complete systems, including sensors, applications and stable elements are uncommon. Some of the approaches designed for protection of IoT networks could be ineffective. Low-power wide-area networks–based protocols can become bizarre if encryption keys are stolen. Virtual private networks are prone to man-in-the-middle attacks.

10.3.2 Physical Tinkering (Tampering)

The edge computing clustered design opens up a fresh, unwanted boundary of practical risk. Although computers and servers that help conventional networks typically are located at large, sometimes extremely protected warehouses, small data centers enabling edge computing might be a security nightmare.

Instead of staying in data centers, these small centers are mostly installed in areas that, as we talk about IoE edge, could be in a corporate office, plant, and anything in between. An intruder who practically tampers with the edge system may bring down a network or damage one of its operators. Saving these systems is away from simple – as they need to be protected against physical threats, it is also a trade-off between protection, expense, and updating and servicing terminal data centers. The device manufacturers will need to be aware of the risks to ensure the protection of sensors to produce local and remote alerts for any tinkering (tampering).

10.3.3 Absence of Security-by-Design Considerations

The primary purpose of edge computing is to provide a more powerful and lightweight computing environment for evolving technologies such as IoE and smart cities. As a result, programmers prefer to concentrate more on performance than on security when building an application-specific edge computing architecture. Such inadequate awareness of security by nature explicitly exposes the edge of computing resources to larger attack surfaces.

10.3.4 Unmigratability of Defense Mechanisms

Security architectures for conventional general-purpose computing systems have been widely studied for quite a long time and are known to be capable of offering solid security assurances in the protection against numerous threats. However, these security architectures cannot be specifically transferred to edge computing networks due to a variety of irresolvable differences, such as opposing computational capacity, multiple open-system environments and apps, separate network topologies, and incompatible protocols. In addition, security technologies developed for one edge computing application may not be explicitly migratable to another scenario for several purposes, such as complexity of edge devices and communication protocols.

10.3.5 Fragmented and Rough-Grained Control

Present access management models for edge computing are inconsistent and rough-grained. They are fractured since different edge computing scenarios follow various access management models that may be configured in radically different ways for segregating, granting, and access permissions. This condition hinders the development of a coherent and manageable access management platform for different edge computing systems. Present access management structures are often coarse-grained as dynamic and under-explored fine-grained permissions are unique to edge computing.

10.3.6 Security in Wireless Communications

Protection of wireless communications is more difficult than protection of standard wired communications. The explanation is that the propagation of a wireless signal is not directed by its source (i.e., free space). When the input is transmitted to the transmitter, the radio signal that reflects it travels in all directions. And where a directional antenna is used, at least a small portion of the signal extends in any direction and can be potentially intercepted.

10.4 Layer-Based Secure Communication Protocols: A Comparison

This section compares layer-based secure communication protocols.

10.4.1 Secure MAC Protocols

Wireless sensor network MAC protocols are elucidated into two different types – time division multiple access (TDMA)-based and carrier sense multiple access (CSMA)-based. TDMA is a centrally managed, coordinated, high-channel usage-based, high-level access protocol. CSMA is a high-channel use-based protocol with a low content, flexible synchronization and random access [3].

TABLE 10.1

Comparison of CSMA MAC Protocols [5]

MAC Protocols	Features	Simulation Environment
IEEE 802.11	Low power, high delay, low throughput	TinyOS (micaz, telosb)
IEEE 802.15.4	Low power, low throughput	TinyOS (micaz, telosb)
S-MAC	High latency, low throughput	TinyOS (micaz) + ns2
T-MAC	Medium latency, low throughput	OMNeT++
DSMAC	Low latency, low throughput	ns2
P-MAC	Low latency, low throughput	ns2

Due to structural variations, conventional MACs cannot be used explicitly in WSNs. As in conventional MAC protocols, quality of operation is the primary target; decreased resource usage is a priority on WSNs [4]. For WSNs various MAC protocols have also been established with different objectives. The comparison of the MAC protocols analyzed is seen in Table 10.1.

IEEE 802.11 [6] is a random withdrawal and carrier signal listening CSMA-based MAC protocol to avoid WSNs data packet collapse. A node that needs to transmit the message, listens to the medium for a limited time and then if the channel is empty, begins to send a message; the protocol also ensures that the channel is reported empty if it fails to detect any contact from nodes. Nodes which lost the contention need to wait and try again. In this procedure, the power-saving mode (PSM) protects the system from hearing prematurely by constantly passing into a dormant condition.

The MAC and physical layers for a wireless private area network (WPAN) are defined in the CSMA/CA protocol, IEEE 802.15.4 [7, 8]. Although this protocol is not expressly configured for WSNs, its low energy consumption, low cost, and simplicity allow it to be used for WNSs. This protocol is currently operating with the Crossbow nodes of Micaz and Telos [9].

The CSMA MAC is the updated IEEE 802.11 CSMA-based MAC protocol [10]. The main purpose is to lower the energy-utilization. The latest characteristics in this protocol are normal listening, colliding elimination, and accidental reception. In the most instances, knots sleep rather than listen to the medium constantly. Hearing and sleep are stable and normal. The synchronization should be strict to allow the nodes to pass together.

In order to be able to transmit large packets more quickly, Sensor-MAC (S-MAC) facilitates message transfer. The nice news about S-MAC is that a TinyOS [11] version of the operating system, in language of nesC [12] running on the model and sensors emulation, greatly decreases energy usage. The drawback of S-MAC is that the nodes require strict synchronization to travel together; it is not coordinated because of stable listening/sleep period, thus raising latency. Time-Out MAC (T-MAC) [13] has been created for WSNs as a CSMA-based MAC. While stable self learning modules (SLMs) improve energy efficiency in S-MACs, they also contribute to high latency and low efficiency. In variable traffic densities, TMAC was proposed for enhancing poor findings of S-MACs. If no conversation takes place in T-MAC (timeout, TA) for any period of time, sleep may occur.

Proper MAC (PMAC) [14] has been built for the WSNs as a CSMA-based MAC protocol. Many MAC protocols, such as S-MAC, frequently sleep to save energy. In these protocols, the duty duration is constant. The sleep-listening periods in PMAC are calculated in a separate manner instead of stable sleeping and listening periods. The traffic of the node and its neighbors decides the pace.

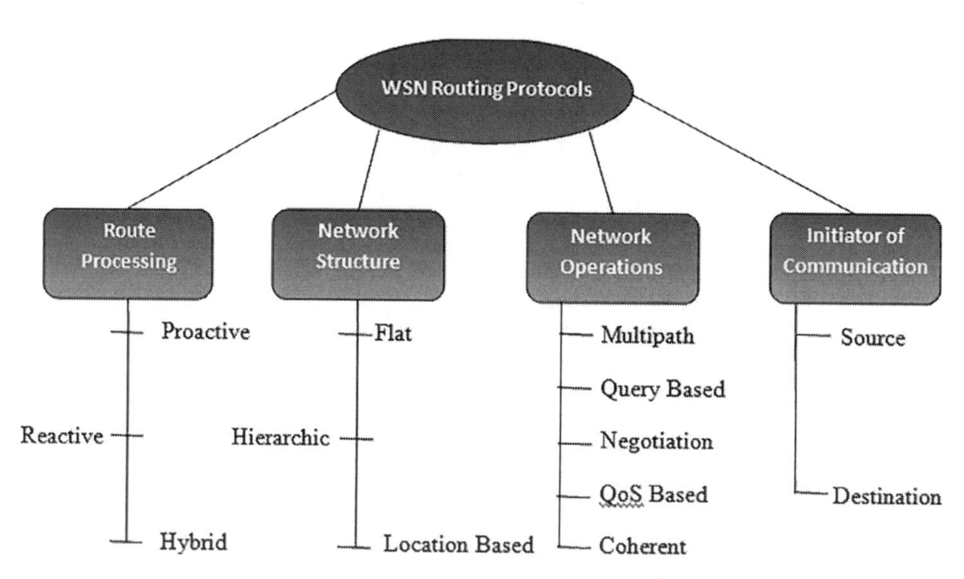

FIGURE 10.4
Categories of routing protocols.

In the S-MAC protocol, Dynamic Sleeping MAC (DSMAC) [15] introduced a dynamic time zone feature. The aim is to reduce latency for applications that are susceptible to delays. All the nodes have one hop latency during the SYNC cycle and start in the same time zone (the elapsed time between the packet in the queue and sending it).

10.4.2 Secure Routing Protocols

Figure 10.4 elucidates different categories of routing protocols.

10.4.2.1 Proactive Routing Protocols

In constructive routing protocol, each node of the network retains a single or several routing tables that are frequently modified. Each node can send a broadcasting message to each node in the network in order to identify network topology change [16].

Open Link State Routing (OLSR) is a constructive [17] routing protocol that is designed for ad hoc mobile networks and can also be used for WSN. The multipoint relay (MRP) system enables the effective flood of control messages across the network. MPR nodes solve the problem of overwhelming the network with control messages. Each node selects MRPs for forwarding control messages that result in the protocol's distributed operation. In addition, a node manages routes constantly for all network destinations such that the protocol is ideal for random and intermittent traffic patterns. This protocol is better suited to traffic dynamics where a large node sub-set connects with other large nodes and the destination and source pairs shift over time. It is primarily ideal for dense and large networks.

10.4.2.2 Reactive Routing Protocols

The reactive routing protocol is an effective on-request protocol for mobile decentralized network. The protocol consists of two primary roles: path exploration and route management. Path exploration is responsible for finding a new route when one is required, and route maintenance is responsible for identifying connection breaks and restoring a current route. By using the incremental search method for route discovery and surroundings repair method for route maintenance, the protocol achieves its bandwidth efficiency. The incremental search approach reduces the number of routing messages so that the number of connections crossed during path discovery is reduced, and thus the bandwidth is used effectively. And by using the surroundings repair process, a node tries to repair its surroundings in case of a connection breakdown, in order to find an alternative path skipping this broken link and thereby attempts to eliminate overhead routing [18].

10.4.2.3 Hybrid Routing Protocol

Hybrid routing protocol has benefited from both the state routing protocols and distance vector protocol and has merged into a new protocol. Usually, these protocols are based on the distance vector protocol but are said to carry much of the characteristics and benefits related to link state routing protocols such as enhanced interior gateway routing protocol (EIGRP) [19].

10.4.2.4 Flat Routing Protocol

A routers network communication protocol, where each router is a partner, is called a flat routing protocol. A flat routing protocol spreads routing information to the connected routers without any segmentation or organization arrangement among them [20].

10.4.2.5 Hierarchical Routing Protocol

The process of grouping routers in a hierarchical way is called hierarchical routing. An example can be the corporate intranet [21]. A high-speed backbone network comprises most corporate intranets. Routers are attached to this backbone and are linked to a specific work group in turn. A special local area network (LAN) occupies these work groups. The reason this is a better arrangement is that the duration (maximum hop count to get from one host to every other host on the network) is two, in spite of hundreds of different work groups being present. Even though a work groups splits their LAN into small segments, in this particular case, the period could only increase to four. The convenience of hierarchical routing is illustrated by taking into consideration the other solutions, with each router connected to another router or each router was connected to two routers.

10.4.2.6 Location-Based Routing

Network size is scalable in location-based routing without raising the overhead of signaling, as routing techniques are essentially localized. Here, through some positioning system like GPS, each node knows its location inside network and this information is used in routing mechanism [22].

10.4.2.7 Multipath Routing

The size of a network is scalable in location-based routing without raising the overhead of signaling, as routing techniques are essentially localized through some positioning system like GPS; every node is knowing its location in network and this information is used in the routing mechanism [23].

10.4.2.8 Query-Based Protocol

Query-driven routing protocols are commonly used in WSNs, in which a sink injects region-of-interest (RoI) queries into the sensor network. Sensor nodes belonging to the RoIs transmit their data towards the sink in response to the query [24].

10.4.2.9 Negotiation Protocol

The negotiation protocol should allow one to negotiate protocols that are used to run multiplexer (MUX) itself beyond the reach of the link. Simple cases should be simple authentication and confidentiality protocol stacks (advertising, negotiation, and use; simple implementation and fair efficiency) [25].

10.4.2.10 QoS-Based Protocol

The network must align energy consumption with service efficiency in quality of service–based routing protocols. (QoS) In particular, when data is transmitted to a sink, certain QoS metrics must be followed, such as delay and bandwidth [26].

10.4.2.11 Coherent Protocol

The nodes perform minimal processing (time stamping and data compression) on data in a coherent data processing protocol before being moved to other sensor nodes or aggregators. Addresses are transferred to a sink node from multiple nodes [27].

10.4.3 Secure Transport-Layer Protocols

Several wireless sensor network transport protocols have been developed to satisfy performance and reliability requirements [28]. The primary feature of the transport layer is reliability, which ensures right information on distribution to the destination or sink node from the source. Reliability mechanisms are available through various protocols suggested. Protocols such as ART, RCRT, ATP, CTCP, STCP, Flush and CRRT [29] provide point-to-point error recovery only for actual error recovery. The destination node will detect and request loss for broadcasting. This will lead to a substantial delay, low flow, and low volume. Other RTMC, CRRT, PSFQ protocols, RMST provides a widespread jump-by-jump error-recovery mechanism. There is no congestion-management mechanism for these RMSTs and PSFQs [29]. The failure of a single packet will not be observed by PSFQ as it does not use Acknowledgement (ACK); rather it uses Negative-ACK (NACK). In comparison, many protocols used NACK (i.e., negative timeout for the stage of loss detection and warning) and used retransmission of the packet for the stage of loss recovery. Each proposed approach has benefits and drawbacks that are suitable for the application. Protocols such as SenTCP, CODA and PCCP don't have any defined process for stability have it for congestion management. In the protocol PCCP, the priority is specified on the point of view of the node instead of the point of view on traffic flow. The distribution of traffic from the node could not be separated. Detection of congestion helps in the detection of potential incidents that may cause network congestion. In order to detect congestion, various protocols use parametric combinations such as packet

TABLE 10.2

List of Different Transport Layer Protocols: A Comparison

Protocol Name	Congestion Detection	Congestion Avoidance	Reliability Level	Type	Reliability Confirmation
PSFQ	–	–	Packet	H-B-H	NACK
ATP	QO	Rate Adjs.	Packet	E-to-E	SACK
RMST	–	–	Packet	H-B-H	NACK
SenTCP	QO, Packet rate	Rate Adjs.	–	–	–
ESRT	QO	Rate Adjs.	Event	E-to-E	–
PCCP	Metric ratio	Rate Adjs.	–	–	–
CODA	QO, Chan. Status	Rate Adjs.	–	–	–
PORT	Node Price	Rate Adjs.	Event	E-to-E	–
ART	ACK to core node	Reduce Traffic of Noncore node	Packet	E-to-E	NACK
STCP	QO	Rate Adjs.	Packet	E-to-E	NACK
CRRT	QO, pkt. Rate	Rate Adjs.	Packet	E-to-E / H-B-H	NACK, ACK
Flush	QO	Rate Adjs.	Packet	E-to-E	NACK
RCRT	Time to recover loss	Rate Adjs.	Packet	E-to-E	NACK Cumm. ACK
CTCP	QO, Trans error loss	Rate Adjs.	Packet	E-to-E	eACK
RTMC	Memory overflow	Header Memory Info	Packet	H-B-H	–

rate, queue occupancy, node price, node latency, link-loss rates, link interruption, ACK obtained to core nodes and time to recover loss, communication error loss, and memory overflow. Table 10.2 justifies the same [30].

The notice of congestion directly or indirectly alerted other nodes. With two common strategies, transport protocols are built with three distinct congestion-avoidance techniques: traffic redirection, rate change, and one seldom-employed mechanism – friendly discussion strategy. Many of them adopt centralized rate adjustment schemes from existing protocols, while Flush, STCP, RTMC and ART use localized schemes. Basic rate adjustment is common approach since, using strict timings, the node actually sends its packet to match the measured rate to apply precise rate adjustment.

10.4.4 Secure Application-Layer Protocols

The application layer protocol in WSN can be classified in three types:

1. Regular applications with broad intervals that characterize those sending low results. We reflect this form of application in NS-2 in a stream of constant bite rate (CBR) with intervals of two seconds. According to a deterministic scale, this CBR stream produces User Datagram Protocol (UDP) traffic. The packets hold a permanent capacity.
2. Highly rated applications describe applications having a broad stream. CBR streams of 0.2 seconds are used in the NS-2 to reflect the network overloading.
3. Applications based on burst during the remaining time, submit information on burst time and sleep, using the Poisson stream of 0.5s on time, 2.5s for off period, and 50kb rate in order to reflect this kind of applications in NS-2. In line with an exponential on/off distribution, the Poisson stream produces traffic. During a time period, packets are sent at a set rate and no packets are sent during off-periods [31].

10.5 Possible Attacks in Intelligent Communication

WSNs are vulnerable and quickly exploited by multiple security threats. The design of deployment, where sensor nodes are distributed without physical guarding duties, is one of the key reasons for these vulnerabilities. That makes the network susceptible to physical attacks [30].

The essence of communication between sensor nodes is another explanation why network communication ranges contain them: intruders can easily send/receive information. Most of the potential WSN attacks are mentioned in this chapter.

10.5.1 Bad-Mouthing Attack

A bad-mouthing attack happens when an attacker attempts to distribute the reputation of the end nodes by communicating these nodes with negative reputation values [33, 34]. A malicious node (A), for example, declares a bad reputation of an innocent node (B). In this case, other sensor nodes will prevent any data from being sent to the node (B), while the node (B) is not the attacker node. If the network has such an attack, the number of isolated nodes will increase for a while and, like all its neighbors, attackers will replicate such actions. The goal of this attack is to separate network nodes to the fullest degree possible.

10.5.2 Good-Mouthing Attack

In this attack, by sending positive reputation values on bad nodes [33, 34], intruders try to deceive the base station or the cluster heads. This attack is the reverse of a bad-mouthing attack. The malicious node (A) has the following form, announcing the positive reputation of another malicious node (B). This attack is intended to dominate the traffic of the network, to break down the entire network.

10.5.3 Whitewashing Attack

A whitewashing attack happens if an infected node attempts to enter the network again with a new identifier and a new reputation [35, 36]. This attack occurs when a malicious node is successfully detected and isolated from the network by the system; then the infected node attempts to reconnect to the network with a fresh identifier to delude the system and have a new value of trust.

10.5.4 Energy-Drain Attack

In this attack, neighbor nodes are asked by a malicious node to respond to useless queries [37, 38]. The infected node typically has infinite power with a high spectrum of communication in order to be able to give its neighbors a number of worthless messages (e.g., controlled messages and compromised data). The goal of the assault is to break down the whole network.

10.5.5 Exhaustion Attack

This attack happens when a malicious node attempts to reenter the network with a new identifier and a new reputation [39, 40]. This attack occurs when a malicious node is successfully detected and isolated from the network by the system; then this malicious node attempts to rejoin the network with a new identifier to delude the system and have a new value of trust.

10.5.6 Homing Attack

Under a homing attack, attackers investigate the traffic on a network in order to understand the geographical area of the base station and cluster heads [40]. They would be able to identify the most critical nodes as intruders know the network configuration and then attack the nodes to easily disrupt the whole network.

10.5.7 Node-Replication Attack

A node-replication attack happens if a duplicate of the node's identifier, which is unique [41], is present. The malicious node occurs in this attack along with an identifier that is allocated to a different node, leading to inappropriate data aggregation. The technique of position estimation, which relies on identifiers of the nodes, will also not be correct.

The case where two different nodes containing the same address are added to the network is shown in Figure 10.5; this is not allowed.

10.5.8 Sybil-Replication Attack

This attack happens when a malicious node has several identities inside the network [42, 43]. Being similar to the replication attack on a server, except with a different unique identifier, the malicious node appears (see Figure 10.6). The Sybil attack can arise in networks where the infected node has more than one A, B, and C identifier. The different standard

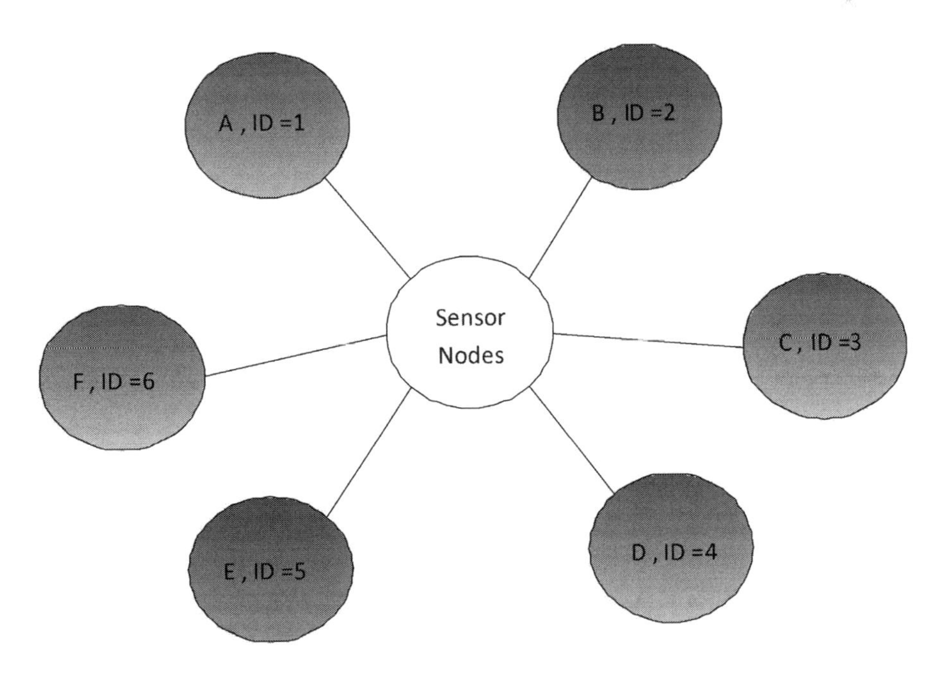

FIGURE 10.5
Node-replication attack.

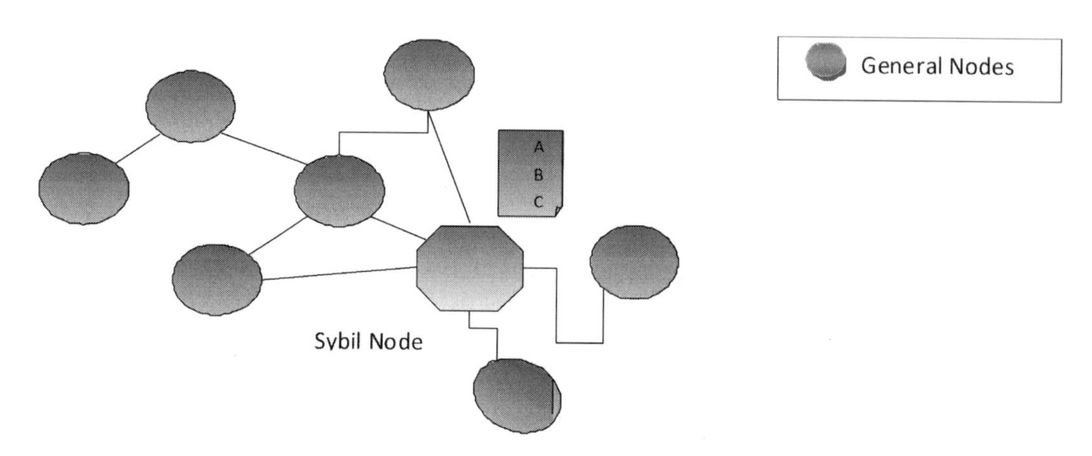

FIGURE 10.6
A scenario of Sybil attack.

node, though, has only one descriptor, such as X, Y, Z nodes, etc. The purpose of the attack is to interrupt the mechanism of data aggregation.

10.6 Issues and Challenges to Design Secure Communication Protocol

There are many challenges in WSN, such as high energy consumption, high bandwidth demand, quality of service (QoS) provisioning, compressing techniques, data processing, and cross-layer design in the physical environment. Nodes with mobility have the ability to compute, sense, and communicate like static nodes [44].

10.6.1 Issues

List of issues is given in Table 10.3.

10.6.2 Challenges in Communication

10.6.2.1 Real Time Challenges

Wireless sensor networks work in environments of the real world. Under certain situations, sensor data should be transmitted within specified time limits in order to be able to

TABLE 10.3
Generic Issues in Communication Protocols

Design Issues	Topology Issues	Other Issues
Fault-tolerant Communication	Geographic Routing	Hardware and OS for WSN
Low Latency	Sensor Holes	Wireless Radio Communication
Scalability	Coverage Topology	Medium Access Schemes
Transmission Media	Adaptability	Deployment & Localization
Coverage Problems		Synchronization & Calibration

make accurate measurements or take action. Till date, very few contributions have been reported in this regard of real-time scenarios in WSNs. Many protocols either neglect real time or actually aim to process as fast as possible and hope the pace is good enough to reach the ends. In real-time network routing, certain outcomes occur. Until now, routing has been the minimum outcome for WSN involving real-time problems. However, WSNs now have enabled other features like data fusion, transfer, aim and identification of event and classification, security and query analysis, and complying with real-time constraints. To ensure soft real-time specifications that engage with the realities of wireless sensor networks (e.g., missed communications, congestion, and noise), new findings are required.

10.6.2.2 Power-Management Challenges

One acclaimed value of WSNs is low-cost practical implementation. Restricted processor bandwidth and limited memory size in WSNs are two major arguable restrictions that will be omitted with fabrication techniques being created. However, because of poor progress in improving battery power, the energy limitation is unlikely to be overcome quickly. In addition, as a practicable alternative, the non-intentional existence of WSN nodes and dangerous sensing conditions prohibit battery replacement. On the other hand, the surveillance nature of many WSN implementations requires a long life, so having energy-efficient surveillance service for a geographical region is a very interesting research topic.

10.6.2.3 Time-Varying and Network-Scaling Characteristics of WSN

The densities of WSNs can differ greatly depending on the application, varying from very sparse to very thick. The behavior of a WSN in these sensor nodes is complex and highly adaptive, as energy forces the WSN need to be self-organized and maintained to continuously change of behavior in response to the current level of operation. In addition, in response to the irregular and volatile behavior of wireless connections induced by high noise levels and radio frequency interference, the sensor nodes can require modification of the behavior to avoid significant performance degradation of the supported application.

10.6.2.4 Management at a Distance

A method of remote management or remote control mechanism can enable administrators to personally control the network deployed at end location.

10.7 Conclusion and Future Scope

A wireless sensor network combined with a neural network may solve large-scale problems in order to improve the capacity to respond to changes in a complex environment or to provide computer intelligence during its operating process after implementation.

Wireless sensor network–based smart ecosystems reflect the next evolutionary phase in engineering growth, such as industrial automation, video detection, traffic management, and robot control. Sensory data comes from numerous interconnected sensor networks with remote locations that are dynamic. In many civilian applications today, WSNs are used. WSNs have received huge input from the high-volume needs of human culture and the development of society.

In order to incorporate intelligence into the implementation of WSNs, many articles propose new concepts. The principle of intelligent WSN implementation for global network adaptation is suggested in this chapter, which adequately solves all possible constraints simultaneously. This chapter offers the modification of multiple parameters to optimize different efficiency metrics such as network lifespan and global energy usage, being dynamically suited to WSN conditions.

For multiple scenarios, multiple experimental effects were shown. This chapter enables WSNs to respond and adjust their functionality according to the requirements of the environment, preserving service efficiency (QoS) and extending the life of the network.

References

1. Jan, B., Farman, H., Javed, H. et al. (2017). Energy Efficient Hierarchical Clustering Approaches in Wireless Sensor Networks: A Survey. Wireless Communications and Mobile Computing 2017, 14. doi: 10.1155/2017/6457942.
2. Rahman, M.A., Saddik, A.E., Gueaieb, W. (2008). Wireless Sensor Network Transport Layer: State of the Art. In: Mukhopadhyay, S., Huang, R. (eds), Sensors. Lecture Notes Electrical Engineering, Vol 21. Springer, Berlin, Heidelberg. doi: 10.1007/978-3-540-69033-7_11.
3. Dener, M., Bay, O.F. (2012). Medium Access Control Protocols for Wireless Sensor Networks: Literature Survey. Gazi University Journal of Science 25(2), 455–464.
4. Ke-Huy, P., Roy, S. (2010). Low-Power Wake-Up Radio for Wireless Sensor Networks. Mobile Networks and Applications 15, 226–236. doi: 10.1007/s11036-009-0184-3.
5. Maitra, T., Roy, S. (2016). A Comparative Study on Popular MAC Protocols for Mixed Wireless Sensor Networks: From Implementation Viewpoint. Computer Science Review 22(C), 107–134. doi: 10.1016/j.cosrev.2016.09.004.
6. "Information Technology – Telecommunications and Information Exchange between Systems – Local and Metropolitan Area Networks – Specific Requirements – Part 11: Wireless LAN Medium Access Control (MAC) and Physical Layer (PHY) Specification," ANSI/IEEE Std 802.11, 1999 Edition.
7. IEEE-TG15.4 (2003). Part 15.4: Wireless Medium Access Control (MAC) and Physical Layer (PHY) Specifications for Low-Rate Wireless Personal Area Networks (LR-WPANs). IEEE Standard for Information Technology.
8. Koubaa, A., Alves, M., Tovar, E. (2007). IEEE 802.15.4: A Federating Communication Protocol for Time-Sensitive Wireless Sensor Networks. In: Sensor Networks and Configurations: Fundamentals, Techniques, Platforms and Experiments. Springer-Verlag, Germany 19–49.
9. Crossbow Technology Inc. (2010). http://www.xbow.com.
10. Ye, W., Heidemann, J., Estrin, D. (2002). "An Energy-Efficient MAC Protocol for Wireless Sensor Networks," Proceedings Twenty-First Annual Joint Conference of the IEEE Computer and Communications Societies, Vol. 3, 1567–1576. New York, NY. doi: 10.1109/INFCOM.2002.1019408.
11. Tiny-OS (2010). http://www.xbow.com.
12. Gay, D., Levis, P., Behren, R. et al., "The nesC Language: A Holistic Approach to Networked Embedded Systems." doi: 10.1145/781131.781133.
13. Dam, T., Langendoen, K. (2003). "An Adaptive Energy-Efficient MAC Protocol for Wireless Sensor Networks," International Conference on Embedded Sensor Networked Sensor Systems, November, 171–180. doi: 10.1145/958491.958512.

14. Zheng, T., Radhakrishna, S., Sarangan, V. (2005). "PMAC: An Adaptive Energy-Efficient MAC Protocol for Wireless Sensor Networks," 19th IEEE International Parallel and Distributed Processing Symposium, 8. Denver, CO. doi: 10.1109/IPDPS.2005.344.
15. Lin, P., Qiao, C., Wang, X. (2004). "Medium Access Control with a Dynamic Duty Cycle for Sensor Networks," 2004 IEEE Wireless Communications and Networking Conference (IEEE Cat. No.04TH8733), Vol. 3, 1534–1539. Atlanta, GA. doi: 10.1109/WCNC.2004.1311671.
16. Biradar, S.R., Sarma, H.K.D., Sarkar, S.K. et al. (2008). "Hybrid (Day-Night) Routing Protocol for Mobile Ad-Hoc Networks," 2008 International Conference on Recent Advances in Microwave Theory and Applications, Jaipur, 875–877. doi: 10.1109/AMTA.2008.4763180.
17. Govindasamy, J., Punniakody, S. (2018). A Comparative Study of Reactive, Proactive and Hybrid Routing Protocol in Wireless Sensor Network Under Wormhole Attack. Journal of Electrical Systems and Information Technology 5(3), 735–744, ISSN 2314-7172.
18. Kalwar, S. (2010). Introduction to Reactive Protocol. IEEE Potentials 29(2), 34–35. doi: 10.1109/MPOT.2009.935243.
19. Dilli, R., Murali Nath, R.S., Shekar Reddy, P.C. (2011) Hybrid Routing for Ad Hoc Wireless Networks. In: Das V.V., Thomas G., Lumban Gaol F. (eds), Information Technology and Mobile Communication. AIM 2011. Communications in Computer and Information Science, Vol. 147. Springer, Berlin, Heidelberg. doi: 10.1007/978-3-642-20573-6_38.
20. Kanavalli, A., Sserubiri, D., Shenoy, P.D. et al. (2009). "A Flat Routing Protocol for Sensor Networks," 2009 Proceeding of International Conference on Methods and Models in Computer Science (ICM2CS), 1–5. Delhi. doi: 10.1109/ICM2CS.2009.5397948.
21. Maitra, T., Barman, S., Giri, D. (2019). Cluster-Based Energy-Efficient Secure Routing in Wireless Sensor Networks. In: Information Technology and Applied Mathematics. Advances in Intelligent Systems and Computing, Vol. 699, 23–40. Springer, Singapore. doi: 10.1007/978-981-10-7590-2_2.
22. Kumar, A., Shwe, H., Wong, K. et al. (2017). Location-Based Routing q Protocols for Wireless Sensor Networks: A Survey. Wireless Sensor Network 9, 25–72. doi: 10.4236/wsn.2017.91003.
23. Gür, G. (2015) Chapter 25 – Multimedia Transmission Over Wireless Networks Fundamentals and Key Challenges. In: Obaidat, M.S., Nicopolitidis, P., Zarai, F. (eds), Modeling and Simulation of Computer Networks and Systems, 717–750. Morgan Kaufmann. ISBN 9780128008874. doi: 10.1016/B978-0-12-800887-4.00025-0.
24. Jain, S., Pattanaik, K.K., Shukla, A. (2019). QWRP: Query-Driven Virtual Wheel Based Routing Protocol for Wireless Sensor Networks with Mobile Sink. Journal of Network and Computer Applications 147, 102430, ISSN 1084-8045. doi: 10.1016/j.jnca.2019.102430.
25. Jin, Y., Geslin, M. (2008). Roles of Negotiation Protocol and Strategy in Collaborative Design. In: Gero, J.S., Goel, A.K. (eds), Design Computing and Cognition '08. Springer, Dordrecht. doi: 10.1007/978-1-4020-8728-8_26.
26. Semchedine, F., Saidi, N.A., Belouzir, L. et al. (2017). QoS-Based Protocol for Routing in Wireless Sensor Networks. Wireless Personal Communications 97, 4413–4429. doi: 10.1007/s11277-017-4731-0.
27. Shabbir, N., Hassan, S. (2017). "Routing Protocols for Wireless Sensor Networks (WSNs)," October. doi: 10.5772/intechopen.70208.
28. Dvir, A., Ta, V.-T., Erlich, S. et al. (2018). STWSN: A Novel Secure Distributed Transport Protocol for Wireless Sensor Networks. International Journal of Communication Systems 31. doi: 10.1002/dac.3827.
29. Sharma, B. and Aseri, Trilok C. (2012). A Comparative Analysis of Reliable and Congestion-Aware Transport Layer Protocols for Wireless Sensor Networks. International Scholarly Research Network ISRN Sensor Networks 2012, 14. doi:10.5402/2012/104057.
30. Karanjawane, A. (2013). Transport Layer Protocol for Urgent Data Transmission in WSN. International Journal of Research in Engineering and Technology 02, 81–89. doi: 10.15623/ijret.2013.0211014.
31. Ramassamy, C., Fouchal, H., Hunel, P. (2012). Impact of Application Layers over Wireless Sensor Networks.

32. Abu Daia, A., Ramadan, R., Fayek, M. (2018). Sensor Networks Attacks Classifications and Mitigation. Annals of Emerging Technologies in Computing 2, 28–43. doi: 10.33166/AETiC.2018.04.003.
33. Shi, E., Perrig, A. (2004). Designing secure sensor networks. IEEE Wireless Communications 11(6), 38–43. doi: 10.1109/MWC.2004.1368895.
34. Thaile, M., Ramanaiah, O.B.V. (2016). "Node Compromise Detection based on NodeTrust in Wireless Sensor Networks," 2016 International Conference on Computer Communication and Informatics (ICCCI), 1–5. Coimbatore. doi: 10.1109/ICCCI.2016.7480020.
35. Collins, M., Dobson, S., Nixon, P. (2008) "A Secure Lightweight Architecture for Wireless Sensor Networks," 2008 The Second International Conference on Mobile Ubiquitous Computing, Systems, Services and Technologies, 349–355. Valencia. doi: 10.1109/UBICOMM.2008.65.
36. Sharma, K., Ghose, M., Yadav, K. (2009). Complete Security Framework for Wireless Sensor Networks. International Journal of Computer Science and Information Security 3(1).
37. Pathan, A.S.K., Lee, H.-W, Hong, C.S. (2006) "Security in Wireless Sensor Networks: Issues and Challenges," 2006 8th International Conference Advanced Communication Technology, 1043–1048. Phoenix Park. doi: 10.1109/ICACT.2006.206151.
38. Umakanth, B., Damodhar, J. (2013) Detection of Energy Draining Attack Using EWMA in Wireless Ad Hoc Sensor Networks. International Journal of Engineering Trends and Technology (IJETT) V4(8), 3691–3695. ISSN:2231-5381.
39. Dubey, A., Jain, V., Kumar, A. (2014). A Survey in Energy Drain Attacks and Their Countermeasures in Wireless Sensor Networks. International Journal of Engineering Research & Technology (IJERT) 3(2), 1206–1210.
40. Znaidi, W., Minier, M., Babau, J.-P. (2008) An Ontology for Attacks in Wireless Sensor Networks. [Research Report] RR-6704, INRIA.
41. Afrand, A., Das, S. (2007). Preventing DoS Attacks in Wireless Sensor Networks: A Repeated Game Theory Approach. International Journal of Network Security, 5(2), 145–153.
42. Mohammadi, S., Ebrahimi Atani, R., Jadidoleslamy, H. (2011). A Comparison of Link Layer Attacks on Wireless Sensor Networks. Journal of Information Security 2, 69–84. doi: 10.4236/jis.2011.22007.
43. Malik, M.Y. (2012). An Outline of Security in Wireless Sensor Networks: Threats, Countermeasures and Implementations. In: Noor Zaman, et al. (eds), Wireless Sensor Networks and Energy Efficiency: Protocols, Routing and Management, 507–527. IGI Global. doi: 10.4018/978-1-4666-0101-7.ch024.
44. Indu, Dixit, S. (2014). Wireless Sensor Networks: Issues & Challenges. International Journal of Computer Science and Mobile Computing, IJCSMC 3(6), 681–685.

Index

Note: Locators in *italics* represent figures and **bold** indicate tables in the text.

S

Z

9780367608897